Statistical Modeling and Machine Learning for Molecular Biology

CHAPMAN & HALL/CRC
Mathematical and Computational Biology Series

Aims and scope:

This series aims to capture new developments and summarize what is known over the entire spectrum of mathematical and computational biology and medicine. It seeks to encourage the integration of mathematical, statistical, and computational methods into biology by publishing a broad range of textbooks, reference works, and handbooks. The titles included in the series are meant to appeal to students, researchers, and professionals in the mathematical, statistical and computational sciences, fundamental biology and bioengineering, as well as interdisciplinary researchers involved in the field. The inclusion of concrete examples and applications, and programming techniques and examples, is highly encouraged.

Series Editors

N. F. Britton
Department of Mathematical Sciences
University of Bath

Xihong Lin
Department of Biostatistics
Harvard University

Nicola Mulder
University of Cape Town
South Africa

Maria Victoria Schneider
European Bioinformatics Institute

Mona Singh
Department of Computer Science
Princeton University

Anna Tramontano
Department of Physics
University of Rome La Sapienza

Proposals for the series should be submitted to one of the series editors above or directly to:
CRC Press, Taylor & Francis Group
3 Park Square, Milton Park
Abingdon, Oxfordshire OX14 4RN
UK

Published Titles

An Introduction to Systems Biology: Design Principles of Biological Circuits
Uri Alon

Glycome Informatics: Methods and Applications
Kiyoko F. Aoki-Kinoshita

Computational Systems Biology of Cancer
Emmanuel Barillot, Laurence Calzone, Philippe Hupé, Jean-Philippe Vert, and Andrei Zinovyev

Python for Bioinformatics
Sebastian Bassi

Quantitative Biology: From Molecular to Cellular Systems
Sebastian Bassi

Methods in Medical Informatics: Fundamentals of Healthcare Programming in Perl, Python, and Ruby
Jules J. Berman

Computational Biology: A Statistical Mechanics Perspective
Ralf Blossey

Game-Theoretical Models in Biology
Mark Broom and Jan Rychtář

Computational and Visualization Techniques for Structural Bioinformatics Using Chimera
Forbes J. Burkowski

Structural Bioinformatics: An Algorithmic Approach
Forbes J. Burkowski

Spatial Ecology
Stephen Cantrell, Chris Cosner, and Shigui Ruan

Cell Mechanics: From Single Scale-Based Models to Multiscale Modeling
Arnaud Chauvière, Luigi Preziosi, and Claude Verdier

Bayesian Phylogenetics: Methods, Algorithms, and Applications
Ming-Hui Chen, Lynn Kuo, and Paul O. Lewis

Statistical Methods for QTL Mapping
Zehua Chen

Normal Mode Analysis: Theory and Applications to Biological and Chemical Systems
Qiang Cui and Ivet Bahar

Kinetic Modelling in Systems Biology
Oleg Demin and Igor Goryanin

Data Analysis Tools for DNA Microarrays
Sorin Draghici

Statistics and Data Analysis for Microarrays Using R and Bioconductor, Second Edition
Sorin Drăghici

Computational Neuroscience: A Comprehensive Approach
Jianfeng Feng

Biological Sequence Analysis Using the SeqAn C++ Library
Andreas Gogol-Döring and Knut Reinert

Gene Expression Studies Using Affymetrix Microarrays
Hinrich Göhlmann and Willem Talloen

Handbook of Hidden Markov Models in Bioinformatics
Martin Gollery

Meta-analysis and Combining Information in Genetics and Genomics
Rudy Guerra and Darlene R. Goldstein

Differential Equations and Mathematical Biology, Second Edition
D.S. Jones, M.J. Plank, and B.D. Sleeman

Knowledge Discovery in Proteomics
Igor Jurisica and Dennis Wigle

Introduction to Proteins: Structure, Function, and Motion
Amit Kessel and Nir Ben-Tal

RNA-seq Data Analysis: A Practical Approach
Eija Korpelainen, Jarno Tuimala, Panu Somervuo, Mikael Huss, and Garry Wong

Introduction to Mathematical Oncology
Yang Kuang, John D. Nagy, and Steffen E. Eikenberry

Biological Computation
Ehud Lamm and Ron Unger

Published Titles (continued)

Chapman & Hall/CRC Mathematical and Computational Biology Series

Statistical Modeling and Machine Learning for Molecular Biology

Alan M. Moses

University of Toronto, Canada

CRC Press
Taylor & Francis Group
Boca Raton   London   New York

CRC Press is an imprint of the
Taylor & Francis Group, an **informa** business

A CHAPMAN & HALL BOOK

CRC Press
Taylor & Francis Group
6000 Broken Sound Parkway NW, Suite 300
Boca Raton, FL 33487-2742

Printed on acid-free paper
Version Date: 20160930

International Standard Book Number-13: 978-1-4822-5859-2 (Paperback)

Library of Congress Cataloging-in-Publication Data

Names: Moses, Alan M., author.
Title: Statistical modeling and machine learning for molecular biology / Alan M. Moses.
Description: Boca Raton : CRC Press, 2016. | Includes bibliographical references and index.
Identifiers: LCCN 2016028358| ISBN 9781482258592 (hardback : alk. paper) | ISBN 9781482258615 (e-book) | ISBN 9781482258622 (e-book) | ISBN 9781482258608 (e-book)
Subjects: LCSH: Molecular biology–Statistical methods. | Molecular biology–Data processing.
Classification: LCC QH506 .M74 2016 | DDC 572.8–dc23
LC record available at https://lccn.loc.gov/2016028358

Visit the Taylor & Francis Web site at
http://www.taylorandfrancis.com

and the CRC Press Web site at
http://www.crcpress.com

For my parents

Contents

SECTION III **Regression**

Acknowledgments

First, I'd like to acknowledge the people who taught me statistics and computers. As with most of the people that will read this book, I took the required semester of statistics as an undergraduate. Little of what I learned proved useful for my scientific career. I came to statistics and computers late, although I learned some html during a high-school job at PCI Geomatics and tried (and failed) to write my first computer program as an undergraduate hoping to volunteer in John Reinitz's lab (then at Mount Sinai in New York). I finally did manage to write some programs as an undergraduate summer student, thanks to Tim Gardner (then a grad student in Marcelo Magnasco's lab), who first showed me PERL codes.

Most of what I learned was during my PhD with Michael Eisen (who reintroduced cluster analysis to molecular biologists with his classic paper in 1998) and postdoctoral work with Richard Durbin (who introduced probabilistic models from computational linguistics to molecular biologists, leading to such universal resources as Pfam, and wrote a classic bioinformatics textbook, to which I am greatly indebted). During my PhD and postdoctoral work, I learned a lot of what is found in this book from Derek Chiang, Audrey Gasch, Justin Fay, Hunter Fraser, Dan Pollard, David Carter, and Avril Coughlan. I was also very fortunate to take courses with Terry Speed, Mark van der Laan, and Michael Jordan while at UC Berkeley and to have sat in on Geoff Hinton's advanced machine learning lectures in Toronto in 2012 before he was whisked off to Google. Most recently, I've been learning from Quaid Morris, with whom I cotaught the course that inspired this book.

I'm also indebted to everyone who read this book and gave me feedback while I was working on it: Miranda Calderon, Drs. Gelila Tilahun, Muluye Liku, and Derek Chiang, my graduate students Mitchell Li Cheong Man, Gavin Douglas, and Alex Lu, as well as an anonymous reviewer.

Much of this book was written while I was on sabbatical in 2014–2015 at Michael Elowitz's lab at Caltech, so I need to acknowledge Michael's generosity to host me and also the University of Toronto for continuing the tradition of academic leave. Michael and Joe Markson introduced me to the ImmGen and single-cell sequence datasets that I used for many of the examples in this book.

Finally, to actually make this book (and the graduate course that inspired it) possible, I took advantage of countless freely available software, R packages, Octave, PERL, bioinformatics databases, *Wikipedia* articles and open-access publications, and supplementary data sets, many of which I have likely neglected to cite. I hereby acknowledge all of the people who make this material available and enable the progress of pedagogy.

I

Overview

The first four chapters give necessary background. The first chapter is background to the book: what it covers and why I wrote it. The next three chapters are background material needed for the statistical modeling and machine learning methods covered in the later chapters. However, although I've presented that material as background, I believe that the review of modeling and statistics (in Chapters 2, 3 and 4) might be valuable to readers, whether or not they intend to go on to the later chapters.

CHAPTER 1

Across Statistical Modeling and Machine Learning on a Shoestring

1.1 ABOUT THIS BOOK

This is a guidebook for biologists about statistics and computers. Much like a travel guide, it's aimed to help intelligent travelers from one place (biology) find their way around a fascinating foreign place (computers and statistics). Like a good travel guide, this book should teach you enough to have an interesting conversation with the locals and to bring back some useful souvenirs and maybe some cool clothes that you can't find at home. I've tried my best to make it fun and interesting to read and put in a few nice pictures to get you excited and help recognize things when you see them.

However, a guidebook is no substitute to having lived in another place—although I can tell you about some of the best foods to try and buildings to visit, these will necessarily only be the highlights. Furthermore, as visitors we'll have to cover some things quite superficially—we can learn enough words to say yes, no, please, and thank you, but we'll never master the language. Maybe after reading the guidebook, some intrepid readers will decide to take a break from the familiar territory of molecular biology for a while and spend a few years in the land of computers and statistics.

Also, this brings up an important disclaimer: A guidebook is not an encyclopedia or a dictionary. This book doesn't have a clear section heading for every topic, useful statistical test, or formula. This means that it won't always be easy to use it for reference. However, because online resources have excellent information about most of the topics covered here, readers are encouraged to look things up as they go along.

1.2 WHAT WILL THIS BOOK COVER?

This book aims to give advanced students in molecular biology enough statistical and computational background to understand (and perform) three of the major tasks of modern machine learning that are widely used in bioinformatics and genomics applications:

1. Clustering

2. Regression

3. Classification

1.2.1 Clustering

Given a set of data, clustering aims to divide the individual observations into groups or clusters. This is a very common problem in several areas of modern molecular biology. In the genomics era, clustering has been applied to genome-wide expression data to find groups of genes with similar expression patterns; it often turns out that these genes do work together (in pathways or networks) in the cell and therefore share common functions. Finding groups of similar genes using molecular interaction data can implicate pathways or help lead to hypotheses about gene function. Clustering has therefore been applied to all types of gene-level molecular interaction data, such as genetic and physical protein interactions. Proteins and genes that share sequence similarity can also be grouped together to delineate "families" that are likely to share biochemical functions. At the other end of the spectrum, finding groups of similar patients (or disease samples) based on molecular profiles is another major current application of clustering.

Historically, biologists wanted to find groups of organisms that represented species. Given a set of measurements of biological traits of individuals, clustering can divide them into groups with some degree of objectivity. In the early days of the molecular era, evolutionary geneticists obtained sequences of DNA and proteins wanting to find patterns that could relate the molecular data to species relationships. Today, inference of population structure by clustering individuals into subpopulations (based on genome-scale genotype data) is a major application of clustering in evolutionary genetics.

Clustering is a classic topic in machine learning because the nature of the groups and the number of groups are unknown. The computer has to "learn" these from the data. There are endless numbers of clustering methods that have been considered, and the bioinformatics literature has contributed a very large number of them.

1.2.2 Regression

Regression aims to model the statistical relationship between one or more variables. For example, regression is a powerful way to test for and model the relationship between genotype and phenotype. Contemporary data analysis methods for genome-wide association studies (GWAS) and quantitative trait loci for gene expression (eQTLs) rely on advanced forms of regression (known as generalized linear mixed models) that can account for complex structure in the data due to the relatedness of individuals and technical biases. Regression methods are used extensively in other areas of biostatistics, particularly in statistical genetics, and are often used in bioinformatics as a means to integrate data for predictive models.

In addition to its wide use in biological data analysis, I believe regression is a key area to focus on in this book for two pedagogical reasons. First, regression deals with the inference of relationships between two or more types of observations, which is a key conceptual issue in all scientific data analysis applications, particularly when one observation can be thought of as predictive or causative of the other. Because classical regression techniques yield straightforward statistical hypothesis tests, regression allows us to connect one type of data to another, and can be used to compare large datasets of different types. Second, regression is an area where the evolution from classical statistics to machine learning methods can be illustrated most easily through the development of penalized likelihood methods. Thus, studying regression can help students understand developments in other areas of machine learning (through analogy with regression), without knowing all the technical details.

1.2.3 Classification

Classification is the task of assigning observations into previously defined classes. It underlies many of the mainstream successes of machine learning: spam filters, face recognition in photos, and the Shazam app. Classification techniques also form the basis for many widely used bioinformatics tools and methodologies. Typical applications include predictions of gene function based on protein sequence or genome-scale experimental data, and identification of disease subtypes and biomarkers. Historically, statistical classification techniques were used to analyze the power of medical tests: given the outcome of a blood test, how accurately could a physician diagnose a disease?

Increasingly, sophisticated machine learning techniques (such as neural networks, random forests and support vector machines or SVMs) are used in popular software for scientific data analysis, and it is essential that modern molecular biologists understand the concepts underlying these. Because of the wide applicability of classification in everyday problems in the information technology industry, it has become a large and rapidly developing area of machine learning. Biomedical applications of these methodological developments often lead to important advances in computational biology. However, before applying these methods, it's critical to understand the specific issues arising in genome-scale analysis, particularly with respect to evaluation of classification performance.

1.3 ORGANIZATION OF THIS BOOK

Chapters 2, 3, and 4 review and introduce mathematical formalism, probability theory, and statistics that are essential to understanding the modeling and machine learning approaches used in contemporary molecular biology. Finally, in Chapters 5 and 6 the first real "machine learning" and nontrivial probabilistic models are introduced. It might sound a bit daunting that three chapters are needed to give the necessary background, but this is the reality of data-rich biology. I have done my best to keep it simple, use clear notation, and avoid tedious calculations. The reality is that analyzing molecular biology data is getting more and more complicated.

You probably already noticed that the book is organized by statistical models and machine learning methods and not by biological examples or experimental data types. Although this makes it hard to look up a statistical method to use on your data, I've organized it this way because I want to highlight the generality of the data analysis methods. For example, clustering can be applied to diverse data from DNA sequences to brain images and can be used to answer questions about protein complexes and cancer subtypes. Although I might not cover your data type or biological question specifically, once you understand the method, I hope it will be relatively straightforward to apply to your data.

Nevertheless, I understand that some readers will want to know that the book covers their type of data, so I've compiled a list of the molecular biology examples that I used to illustrate methods.

LIST OF MOLECULAR BIOLOGY EXAMPLES

1. Chapter 2—Single-cell RNA-seq data defies standard models
2. Chapter 2—Comparing RNA expression between cell types for one or two genes
3. Chapter 2—Analyzing the number of kinase substrates in a list of genes
4. Chapter 3—Are the genes that came out of a genetic screen involved in angiogenesis?
5. Chapter 3—How many genes have different expression levels in T cells?
6. Chapter 3—Identifying eQTLs
7. Chapter 4—Correlation between expression levels of CD8 antigen alpha and beta chains
8. Chapter 4—GC content differences on human sex chromosomes
9. Chapter 5—Groups of genes and cell types in the immune system
10. Chapter 5—Building a tree of DNA or protein sequences
11. Chapter 5—Immune cells expressing CD4, CD8 or both
12. Chapter 5—Identifying orthologs with OrthoMCL
13. Chapter 5—Protein complexes in protein interaction networks
14. Chapter 6—Single-cell RNA-seq revisited
15. Chapter 6—Motif finding with MEME
16. Chapter 6—Estimating transcript abundance with Cufflinks
17. Chapter 6—Integrating DNA sequence motifs and gene expression data
18. Chapter 7—Identifying eQTLs revisited
19. Chapter 7—Does mRNA abundance explain protein abundance?
20. Chapter 8—SAG1 expression is controlled by multiple loci
21. Chapter 8—mRNA abundance, codon bias, and the rate of protein evolution
22. Chapter 8—Predicting gene expression from transcription factor binding motifs
23. Chapter 8—Motif finding with REDUCE
24. Chapter 9—Modeling a gene expression time course

25. Chapter 9—Inferring population structure with STRUCTURE
26. Chapter 10—Are mutations harmful or benign?
27. Chapter 10—Finding a gene expression signature for T cells
28. Chapter 10—Identifying motif matches in DNA sequences
29. Chapter 11—Predicting protein folds
30. Chapter 12—The BLAST homology detection problem
31. Chapter 12—LPS stimulation in single-cell RNA-seq data

1.4 WHY ARE THERE MATHEMATICAL CALCULATIONS IN THE BOOK?

Although most molecular biologists don't (and don't want to) do mathematical derivations of the type that I present in this book, I have included quite a few of these calculations in the early chapters. There are several reasons for this. First of all, the type of machine learning methods presented here are mostly based on probabilistic models. This means that the methods described here really are mathematical things, and I don't want to "hide" the mathematical "guts" of these methods. One purpose of this book is to empower biologists to unpack the algorithms and mathematical notations that are buried in the methods section of most of the sophisticated primary research papers in the top journals today. Another purpose is that I hope, after seeing the worked example derivations for the classic models in this book, some ambitious students will take the plunge and learn to derive their own probabilistic machine learning models. This is another empowering skill, as it frees students from the confines of the prepackaged software that everyone else is using. Finally, there are students out there for whom doing some calculus and linear algebra will actually be fun! I hope these students enjoy the calculations here. Although calculus and basic linear algebra are requirements for medical school and graduate school in the life sciences, students rarely get to use them.

I'm aware that the mathematical parts of this book will be unfamiliar for many biology students. I have tried to include very basic introductory material to help students feel confident interpreting and attacking equations. This brings me to an important point: although I don't assume any prior knowledge of statistics, I do assume that readers are familiar with multivariate calculus and something about linear algebra (although I do review the latter briefly). But don't worry if you are a little rusty and don't remember, for example, what a partial derivative is; a quick visit to Wikipedia might be all you need.

A PRACTICAL GUIDE TO ATTACKING A MATHEMATICAL FORMULA

For readers who are not used to (or afraid of) mathematical formulas, the first thing to understand is that unlike the text of this book, where I try to explain things as directly as possible, the mathematical formulas work differently. Mathematical knowledge has been suggested to be a different kind of knowledge, in that it reveals itself to each of us as we come to "understand" the formulas (interested readers can refer to Heidegger on this point). The upshot is that to be understood, formulas must be contemplated quite aggressively— hence they are not really read, so much as "attacked." If you are victorious, you can expect a good formula to yield a surprising nugget of mathematical truth. Unlike normal reading, which is usually done alone (and in one's head) the formulas in this book are best attacked out loud, rewritten by hand, and in groups of 2 or 3.

When confronted with a formula, the first step is to make sure you know what the point of the formula is: What do the symbols mean? Is it an equation (two formulas separated by an equals sign)? If so, what kind of a thing is supposed to be equal to what? The next step is to try to imagine what the symbols "really" are. For example, if the big "sigma" (that means a sum) appears, try to imagine some examples of the numbers that are in the sum. Write out a few terms if you can. Similarly, if there are variables (e.g., x) try to make sure you can imagine the numbers (or whatever) that x is trying to represent. If there are functions, try to imagine their shapes. Once you feel like you have some understanding of what the formula is trying to say, to fully appreciate it, a great practice is to try using it in a few cases and see if what you get makes sense. What happens as certain symbols reach their limits (e.g., become very large or very small)?

For example, let's consider the Poisson distribution:

$$P(X|\lambda) = \frac{e^{-\lambda}\lambda^X}{X!} \quad \lambda > 0, \ X \in \{0,1,2,...\}$$

First of all, there are actually three formulas here. The main one on the left, and two small ones on the right. Let's start with the small ones. The first part is a requirement that λ is a positive number. The other part tells us what X is. I have used fancy "set" notation that says "X is a member of the set that contains the numbers 0, 1, 2 and onwards until infinity." This means X can take on one of those numbers.

The main formula is an equation (it has an equals sign) and it is a function—you can get this because there is a letter with parentheses next to it, and the parentheses are around symbols that reappear on the right. The function is named with a big "P" in this case, and there's a "|" symbol inside the parentheses. As we will discuss in Chapter 2, from seeing these two together, you can guess that the "P" stands for probability, and the "|" symbol refers to conditional probability. So the formula is giving an equation for the conditional probability of X given λ. Since we've guessed that the equation is a probability distribution, we know that X is a random variable, again discussed in Chapter 2, but for our purposes now, it's something that can be a number.

Okay, so the formula is a function that gives the probability of X. So what does the function look like? First, we see an "e" to the power of negative λ. e is just a special number (a fundamental constant, around 2.7) and λ is another positive number whose value is set to be something greater than 0. Any number to a negative power gets very small as the exponent gets big, and goes to 1 when the exponent goes to 0. So this first part is just a number that doesn't depend on X. On the bottom, there's an $X!$ The factorial sign means $a! = a \times (a-1) \times (a-2) \times \cdots \times (2 \times 1)$, which will get big "very" fast as X gets big. However, there's also a λ^X which will also get very big, very fast if λ is more than 1. If λ is less than 1, λ^X, will get very small, very fast as X gets big. In fact, if λ is less than 1, the $X!$ will dominate the formula, and the probability will simply get smaller and smaller as X gets bigger (Figure 1.1, left panel). As λ approaches 0, the formula approaches 1 for $X=0$ (because any number to the power of 0 is still 1, and 0! is defined to be 1) and 0 for everything else (because a number approaching zero to any power is still 0, so the formula will have a 0 in it, no matter what the value of X). Not too interesting. If λ is more than 1, things get a bit more interesting, as there will be a competition between λ^X and $X!$ The e term will just get smaller. It turns out that factorials grow faster than exponentials (Figure 1.1, right panel), so the bottom will always end up bigger than the top, but this is not something that would be obvious, and for intermediate values of X, the exponential might be bigger (e.g., $3! = 6 < 2^3 = 8$).

Another interesting thing to note about this formula is that for $X=0$ the formula is always just $e^{-\lambda}$ and for $X=1$, it's always $\lambda e^{-\lambda}$. These are

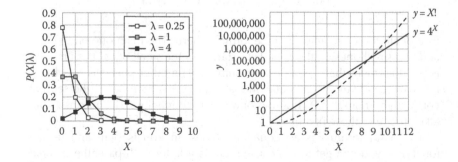

FIGURE 1.1 Graphs illustrating some things about the formula for the Poisson distribution. The left panel shows the value of the formula for different choices of λ. On the right is the competition between λ^X and $X!$ for $\lambda=4$. Note that the y-axis is in log scale.

equal when $\lambda = 1$, which means that the probability of seeing 0 is equal to the probability of seeing 1 only when $\lambda = 1$, and that probability turns out to be $1/e$.

So I went a bit overboard there, and you probably shouldn't contemplate that much when you encounter a new formula—those are, in fact, thoughts I've had about the Poisson distribution over many years. But I hope this gives you some sense of the kinds of things you can think about when you see a formula.

1.5 WHAT WON'T THIS BOOK COVER?

Despite several clear requests to include them, I have resisted putting R, python, MATLAB®, PERL, or other code examples in the book. There are two major reasons for this. First, the syntax, packages, and specific implementations of data analysis methods change rapidly—much faster than the foundational statistical and computational concepts that I hope readers will learn from this book. Omitting specific examples of codes will help prevent the book from becoming obsolete by the time it is published. Second, because the packages for scientific data analysis evolve rapidly, figuring out how to use them (based on the accompanying user manuals) is a key skill for students. This is something that I believe has to be learned through experience and experimentation—as the PERL mantra goes, "TMTOWTDI: there's more than one way to do it"—and while code examples might speed up research in the short term, reliance on them hinders the self-teaching process.

Sadly, I can't begin to cover all the beautiful examples of statistical modeling and machine learning in molecular biology in this book. Rather, I want to help people understand these techniques better so they can go forth and produce more of these beautiful examples. The work cited here represents a few of the formative papers that I've encountered over the years and should not be considered a review of current literature. In focusing the book on applications of clustering, regression, and classification, I've really only managed to cover the "basics" of machine learning. Although I touch on them briefly, hidden Markov models or HMMs, Bayesian networks, and deep learning are more "advanced" models widely used in genomics and bioinformatics that I haven't managed to cover here. Luckily, there are more advanced textbooks (mentioned later) that cover these topics with more appropriate levels of detail.

The book assumes a strong background in molecular biology. I won't review DNA, RNA proteins, etc., or the increasingly sophisticated, systematic experimental techniques used to interrogate them. In teaching this material to graduate students, I've come to realize that not all molecular biology students will be familiar with all types of complex datasets, so I will do my best to introduce them briefly. However, readers may need to familiarize themselves with some of the molecular biology examples discussed.

1.6 WHY IS THIS A BOOK?

I've been asked many times by students and colleagues: Can you recommend a book where I can learn the statistics that I need for bioinformatics and genomics? I've never been able to recommend one. Of course, current graduate students have access to all the statistics and machine learning reference material they could ever need via the Internet. However, most of it is written for a different audience and is Impenetrable to molecular biologists. So, although all the formulas and algorithms in this book are probably easy to find on the Internet, I hope the book format will give me a chance to explain in simple and accessible language what it all means.

Historically speaking, it's ironic that contemporary biologists should need a book to explain data analysis and statistics. Much of the foundational work in statistics was developed by Fisher, Pearson, and others out of direct need to analyze biological observations. With the ascendancy of digital data collection and powerful computers, to say that data analysis has been revolutionized is a severe understatement at best. It is simply not possible for biologists to keep up with the developments in statistics and computer science that are introducing ever new and sophisticated computer-enabled data analysis methods.

My goal is that the reader will be able to situate their molecular biology data (ideally that results from the experiments they have done) in relation to analysis and modeling approaches that will allow them to ask and answer the questions in which they are most interested. This means that if the data really is just two lists of numbers (say, for mutant and wt) they will realize that all they need is a t-test, (or a nonparametric alternative if the data are badly behaved.)

In most practical cases, however, the kinds of questions that molecular biologists are asking go far beyond telling if mutant is different than wild type. In the information age, students need to quantitatively integrate their data with other datasets that have been made publically available;

they may have done several types of experiments that need to be combined in a rigorous framework.

This means that, ideally, a reader of this book will be able to understand the sophisticated statistical approaches that have been applied to their problem (even if they are not covered explicitly in this book) and, if necessary, they will have the tools and context to develop their own statistical model or simple machine learning method.

As a graduate student in the early 00s, I also asked my professors for books, and I was referred (by Terry Speed, a statistical geneticist Dudoit (2012)) to a multivariate text book by Mardia, Kent, and Bibby, which I recommend to anyone who wants to learn multivariate statistics. It was at that time I first began to see statistics as more than an esoteric collection of strange "tests" named after long-dead men. However, Mardia et al. is from the 1980, and is out of date for modern molecular biology applications. Similarly, I have a copy of Feller's classic book that my PhD supervisor Mike Eisen once gave to me, but this book really isn't aimed at molecular biologists—P-value isn't even in the index of Feller. I still can't recommend a book that explains what a P-value is in the context of molecular biology. Books are either way too advanced for biologists (e.g., Friedman, Tibshirani, and Hastie's *The Elements of Statistical Learning* or MacKay's *Information Theory, Inference, and Learning Algorithms*), or they are out of date with respect to modern applications. To me the most useful book is *Biological Sequence Analysis: Probabilistic Models of Proteins and Nucleic Acids* by Durbin et al. (1998). Although that book is focused on bioinformatics, I find the mix of theory and application in that book exceptionally useful—so much so that it is still the book that I (and my graduate students) read 15 years later. I am greatly indebted to that book, and I would strongly recommend it to anyone who wants to understand HMMs.

In 2010, Quaid Morris and I started teaching a short course called "ML4Bio: statistical modeling and machine learning for molecular biology" to help our graduate students get a handle on data analysis. As I write this book in 2016, it seems to me that being able to do advanced statistical analysis of large datasets is the most valuable transferrable skill that we are teaching our bioinformatics students. In industry, "data scientists" are tasked with supporting key business decisions and get paid big $$$. In academia, people who can formulate and test hypotheses on large datasets are leading the transformation of biology to a data-rich science.

REFERENCES AND FURTHER READING

Dudoit S. (Ed.). (2012). *Selected Works of Terry Speed*. New York: Springer.

Durbin R, Eddy SR, Krogh A, Mitchison G. (1998). *Biological Sequence Analysis: Probabilistic Models of Proteins and Nucleic Acids*, 1st edn. Cambridge, U.K.: Cambridge University Press.

Feller W. (1968). *An Introduction to Probability Theory and Its Applications*, Vol. 1, 3rd edn. New York: Wiley.

Hastie T, Tibshirani R, Friedman J. (2009). *The Elements of Statistical Learning*. New York: Springer.

MacKay DJC. (2003). *Information Theory, Inference, and Learning Algorithms*, 1st edn. Cambridge, U.K.: Cambridge University Press.

Mardia K, Kent J, Bibby J. (1980). *Multivariate Analysis*, 1st edn. London, U.K.: Academic Press.

CHAPTER 2

Statistical Modeling

S O YOU'VE DONE AN experiment. Most likely, you've obtained numbers. If you didn't, you don't need to read this book. If you're still reading, it means you have some numbers—data. It turns out that data are no good on their own. They need to be *analyzed*. Over the course of this book, I hope to convince you that the best way to think about analyzing your data is with statistical modeling. Even if you don't make it through the book, and you don't want to ever think about models again, you will still almost certainly find yourself using them when you analyze your data.

2.1 WHAT IS STATISTICAL MODELING?

I think it's important to make sure we are all starting at the same place. Therefore, before trying to explain statistical modeling, I first want to discuss just *plain* modeling.

Modeling (for the purposes of this book) is the attempt to describe a series of measurements (or other kinds of numbers or events) using mathematics. From the perspective of machine learning, a model can only be considered useful if it can describe the series of measurements more succinctly (or compactly) than the list of numbers themselves. Indeed, in one particularly elegant formulation, the information the machine "learns" is precisely the difference between the length of the list of numbers and its compact representation in the model. However, another important thing (in my opinion) to ask about a model, besides its compactness, is whether it provides some kind of "insight" or "conceptual simplification" about the numbers in question.

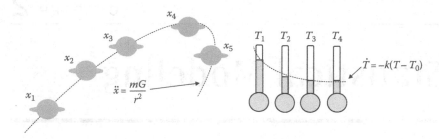

FIGURE 2.1 Example of observations explained by Newton's law of universal gravitation and Newton's law of cooling. Models attempt to describe a series of numbers using mathematics. On the left, observations $(x_1, x_2, x_3, \ldots)$ of the position of a planet as it wanders through the night sky are predicted (dotted line) by Newton's law of universal gravitation. On the right, thermometer readings $(T_1, T_2, T_3, \ldots)$ decreasing according to Newton's law of cooling describe the equilibration of the temperature of an object whose temperature is greater than its surroundings (T_0).

Let's consider a very simple example of a familiar model: Newton's law of universal gravitation (Figure 2.1, left panel). A series of measurements of a flying (or falling) object can be replaced by the starting position and velocity of the object, along with a simple mathematical formula (second derivative of the position is proportional to mass over distance squared), and some common parameters that are shared for most flying (or falling) objects. What's so impressive about this model is that (1) it can predict a *huge* number of subsequent observations, with only a few parameters, and (2) it introduces the concept of gravitational force, which helps us understand why things move around the way they do.

It's important to note that physicists do not often call their models "models," but, rather "laws." This is probably for either historical or marketing reasons ("laws of nature" sounds more convincing than "models of nature"), but as far as I can tell, there's no difference *in principle* between Newton's law of universal gravitation and any old model that we might see in this book. In practice, the models we'll make for biology will probably not have the simplicity or explanatory power that Newton's laws or Schrodinger's equation have, but that is a difference of degree and not kind. Biological models are probably more similar to a different, lesser known physics model, also attributed to Newton: his (nonuniversal) "law of cooling," which predicts measurements of the temperatures of certain types of hot things as they cool off (Figure 2.1, right panel). Although this model doesn't apply to all hot things, once you've found a hot thing that does fit this model, you can predict the temperature of the hot thing over time, simply based on the difference between the temperature of the thing and its surroundings. Once again, a simple mathematical formula predicts many observations, and we have the simple insight that the rate at which objects cool is simply proportional to how much hotter they are than the surroundings. Much like this "law of cooling," once we've identified a biological system that we can explain using a simple mathematical formula, we can compactly represent the behavior of that system.

We now turn to statistical modeling, which is our point here. Statistical modeling also tries to represent some observations of numbers or events—now called a "sample" or "data"—in a more compact way, but it includes the possibility of randomness in the observations. Without getting into a philosophical discussion on what this "randomness" is (see my next book), let's just say that statistical models acknowledge that the data will not be "fully" explained by the model. Statistical models will be happy to predict *something* about the data, but they will not be able to precisely reproduce the exact list of numbers. One might say, therefore, that statistical models are inferior, because they don't have the explaining power that, say, Newton's laws have, because Newton always gives you an exact answer. However, this assumes that you *want* to explain your data exactly; in biology, you almost never do. Every biologist knows that whenever they write down a number, a part of the observation is actually just randomness or noise due to methodological, biological, or other experimental contingencies. Indeed, it was a geneticist (Fisher) who really invented statistics after all—250 years after Newton's law of universal gravitation. Especially with the advent of high-throughput molecular biology, it has never been more true that much of what we measure in biological experiments is noise or randomness. We spend a lot of time and energy measuring things that we can't and don't really want to explain. That's why we need statistical modeling.

Although I won't spend more time on it in this book, it's worth noting that sometimes the randomness in our biological observations is interesting and is something that we do want to explain. Perhaps this is most well-appreciated in the gene expression field, where it's thought that inherently stochastic molecular processes create inherent "noise" or stochasticity in gene expression (McAdams and Arkin 1997). In this case, there has even been considerable success predicting the mathematical form of the variability based on biophysical assumptions (Shahrezaei and Swain 2008). Thus, statistical modeling is not only a convenient way to deal with imperfect experimental measurements, but, in some cases, the only way to deal with the inherent stochasticity of nature.

2.2 PROBABILITY DISTRIBUTIONS ARE THE MODELS

Luckily for Fisher, randomness had been studied for many years, because (it turns out) people can get a big thrill out of random processes—dice and cards—especially if there's money or honor involved. Beautiful mathematical models of randomness can be applied to model the part of biological data that can't be (or isn't interesting enough to be) explained. Importantly, in so doing, we will (hopefully) separate out the interesting part.

A very important concept in statistical modeling is that any set of data that we are considering could be "independent and identically distributed" or "i.i.d." for short, which means that all of the measurements in the dataset can be thought of as coming from the same "pool" of measurements, so that the first observation could have just as easily been the seventh observation. In this sense, the observations are identically distributed. Also, the observations don't know or care about what other observations have been made before (or will be made after) them. The eighth coin toss is just as likely to come up "heads," even if the first seven were also heads. In this sense, the observations are independent. In general, for data to be well-described by a simple mathematical model, we will want our data to be i.i.d., and this assumption will be made implicitly from now on, unless stated otherwise.

As a first example of a statistical model, let's consider the lengths of iris petals (Fisher 1936), very much like what Fisher was thinking about while he was inventing statistics in the first place (Figure 2.2). Since iris petals can be measured to arbitrary precision, we can treat these as so-called "real" numbers, namely, numbers that can have decimal points or fractions, etc. One favorite mathematical formula that we can use to describe the randomness of real numbers is the Gaussian, also called the normal distribution, because it appears so ubiquitously that it is simply "normal" to find real numbers with Gaussian distributions in nature.

More generally, this example shows that when we think about statistical modeling, we are not trying to explain exactly what the observations will be: The iris petals are considered to be i.i.d., so if we measure another

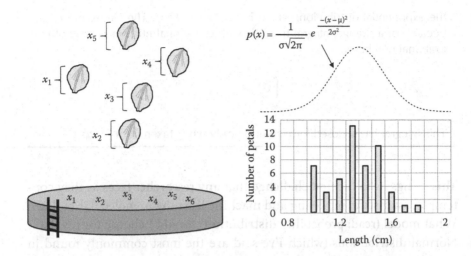

$$p(x) = \frac{1}{\sigma\sqrt{2\pi}} e^{\frac{-(x-\mu)^2}{2\sigma^2}}$$

FIGURE 2.2 The simplest statistical models. Left: Measurements of petals' sizes from Fisher's iris data can be thought of as observations in a pool. Top right is the formula and the predicted "bell curve" for the Gaussian probability distribution, while the bottom right shows the numbers of petals of *I. versicolor* within each size bin ($n = 50$ petals total).

sample of petals, we will not observe the same numbers again. The statistical model tries to say something useful about the sample of numbers, without trying to say exactly what those numbers will be. The Gaussian distribution describes (quantitatively) the way the numbers will tend to behave. This is the essence of statistical modeling.

DEEP THOUGHTS ABOUT THE GAUSSIAN DISTRIBUTION

The distribution was probably first introduced as an easier way to calculate binomial probabilities, which were needed for predicting the outcome of dice games, which were very popular even back in 1738. I do not find it obvious that there should be any mathematical connection between predictions about dice games and the shapes of iris petals. It is a quite amazing fact of nature that the normal distribution describes both very well. Although for the purposes of this book, we can just consider this an empirical fact of nature, it is thought that the universality of the normal distribution in nature arises due to the "central limit theorem" that governs the behavior of large collections of random numbers. We shall return to the central limit theorem later in this chapter.

Another amazing thing about the Gaussian distribution is its strange formula. It's surprising to me that a distribution as "normal" as the Gaussian would have such a strange looking formula (Figure 2.2). Compare it to, say, the exponential distribution, which is just $p(x) = \lambda e^{-\lambda x}$. The Gaussian works because of a strange integral that relates the irrational number e, to another irrational number π.

$$\int_{-\infty}^{\infty} e^{-x^2} dx = \sqrt{\pi}$$

This integral has a closed form solution only when taken from $-\infty$ to $+\infty$.

This brings us to the first challenge that any researcher faces as they contemplate statistical modeling and machine learning in molecular biology: What model (read: probability distribution) should I choose for my data? Normal distributions (which I've said are the most commonly found in nature) are defined for real numbers like -6.72, 4.300, etc. If you have numbers like these, consider yourself lucky because you might be able to use the Gaussian distribution as your model. Molecular biology data comes in many different types, and unfortunately much of it doesn't follow a Gaussian distribution very well. Sometimes, it's possible to transform the numbers to make the data a bit closer to Gaussian. The most common way to do this is to try taking the logarithm. If your data are strictly positive numbers, this might make the distribution more symmetric. And taking the logarithm will not change the relative order of datapoints.

In molecular biology, there are three other major types of data that one typically encounters (in addition to "real numbers" that are sometimes well-described by Gaussian distributions). First, "categorical" data describes data that is really not well-described by a sequence of numbers, such as experiments that give "yes" or "no" answers, or molecular data, such as genome sequences, that can be "A," "C," "G," or "T." Second, "fraction" or "ratio" data is when the observations are like a collection of yes or no answers, such 13 out of 5212 or 73 As and 41 Cs. Finally, "ordinal" data is when data is drawn from the so-called "natural" numbers, 0, 1, 2, 3, ... In this case, it's important that 2 is more than 1, but it's not possible to observe anything in between.

Depending on the data that the experiment produces, it will be necessary to choose appropriate probability distributions to model it. In general, you can test whether your data "fits" a certain model by graphically comparing the distribution of the data to the distribution predicted by statistical theory. A nice way to do this is using a quantile–quantile plot or "qq-plot" (Figure 2.3). If they don't disagree too badly, you can usually

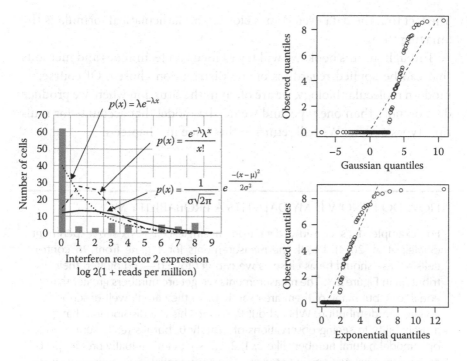

FIGURE 2.3 Modeling single-cell RNA-seq data (Shalek et al. 2014). The histogram on the left shows the number of cells with each expression level indicated with gray bars. The predictions of three probability distributions are shown as lines and are indicated by their formulas: The exponential, Poisson and Gaussian distributions are shown from top to bottom. The data was modeled in log-space with 1 added to each observation to avoid log(0), and parameters for each distribution were the maximum likelihood estimates. On the right are "quantile-quantile" plots comparing the predicted quantiles of the probability distribution to the observations. If the data fit the distributions, the points would fall on a straight line. The top right plot shows that the fit to a Gaussian distribution is very poor: Rather than negative numbers (expression levels less than 1), the observed data has many observations that are exactly 0. In addition, there are too many observations of large numbers. The exponential distribution, shown on the bottom right fits quite a bit better, but still, greatly underestimates the number of observations at 0.

be safe assuming your data are consistent with that distribution. It's important to remember that with large genomics datasets, you will likely have enough power to reject the hypothesis that your data "truly" come from any distribution. Remember, there's no reason that your experiment should generate data that follows some mathematical formula.

The fact that the data even comes close to a mathematical formula is the amazing thing.

Throughout this book, we will try to focus on techniques and methods that can be applied regardless of the distribution chosen. Of course, in modern molecular biology, we are often in the situation where we produce data of more than one type, and we need a model that accounts for multiple types of data. We will return to this issue in Chapter 4.

HOW DO I KNOW IF MY DATA FITS A PROBABILITY MODEL?

For example, let's consider data from a single-cell RNA-seq experiment (Shalek et al. 2014). I took the measurements for 1 gene from 96 control cells—these should be as close as we can get to i.i.d—and plotted their distribution in Figure 2.3. The measurements we get are numbers greater than or equal to 0, but many of them are exactly 0, so they aren't well-described by a Gaussian distribution. What about Poisson? This is a distribution that gives a probability to seeing observations of exactly 0, but it's really supposed to only model natural numbers like 0, 1, 2, ..., so it can't actually predict probabilities for observations in between. The exponential is a continuous distribution that is strictly positive, so it is another candidate for this data.

Attempts to fit this data with these standard distributions are shown in Figure 2.3. I hope it's clear that all of these models underpredict the large number of cells with low expression levels, (exactly 0) as well as the cells with very large expression levels. This example illustrates a common problem in modeling genomic data. It doesn't fit very well with any of the standard, simplest models used in statistics. This motivates the use of fancier models, for example, the data from single-cell RNA-seq experiments can be modeled using a mixture model (which we will meet in Chapter 5).

But how do I know the data don't fit the standard models? So far, I plotted the histogram of the data compared to the predicted probability distribution and argued that they don't fit well based on the plot. One can make this more formal using a so-called quantile–quantile plot (or qq-plot for short). The idea of the qq-plot is to compare the amount of data appearing up to that point in the distribution to what would be expected based on the mathematical formula for the distribution (the theory). For example, the Gaussian distribution predicts that the first ~0.2% of the data falls below 3 standard deviations of the mean, 2.5% of the data falls below 2 standard deviations of the mean, 16% of the data falls below 1 standard deviation of the mean, etc. For any observed set of data, it's possible to calculate these so-called "quantiles" and compare them to the predictions of the Gaussian (or any standard) distribution. The qq-plot compares the amount of data we expect in a certain part of the range to what was actually observed. If the quantiles of the observed data agree pretty well with the quantiles that we expect, we will see a straight line on the qq-plot. If you are doing a statistical test that depends on the assumption that your data follows a certain distribution, it's a quick and easy check to make a qq-plot using *R* and see how reasonable the assumption is.

Perhaps more important than the fit to the data is that the models make very different qualitative claims about the data. For example, the Poisson model predicts that there's a typical expression level we expect (in this case, around 3), and we expect to see fewer cells with expression levels much greater or less than that. On the other hand, the exponential model predicts that many of the cells will have 0, and that the number of cells with expression above that will decrease monotonically as the expression level gets larger. By choosing to describe our data with one model or the other, we are making a very different decision about what's important.

2.3 AXIOMS OF PROBABILITY AND THEIR CONSEQUENCES: "RULES OF PROBABILITY"

More generally, probability distributions are mathematical formulas that assign to events (or more technically observations of events) numbers between 0 and 1. These numbers tell us something about the relative rarity of the events. This brings us to my first "rule of probability": For two mutually exclusive events, the sum of their probabilities is the probability of one *or* the other happening. From this rule, we can already infer another rule, which is that under a valid probability distribution, that the sum of *all* possible observations had better be exactly 1, because *something* has to happen, but it's not possible to observe *more* than one event in each try. (Although it's a bit confusing, observing "2" or "−7.62" of something would be considered one event for our purposes here. In Chapter 4 we will consider probability distributions that model multiple simultaneous events known as multivariate distributions.)

The next important rule of probability is the "joint probability," which is the probability of a series of independent events happening: The joint probability is the product of the individual probabilities. The last and possibly most important rule of probability is about conditional probabilities. Conditional probability is the probability of an event *given* that another event already happened. It is the joint probability divided by the probability of the event that already happened.

Intuitively, if two events are independent, then the conditional probability should be the same as the joint probability: If X and Y are independent, then X shouldn't care if Y has happened or not. The probability of

X given Y should just be the probability of X. As an exercise, let's see if we can use the rules of probability to show this mathematically. Starting with the definition of the conditional probability,

$$P(X = A | Y = B) = \frac{P(X = A, Y = B)}{P(Y = B)}.$$

Since in this example, the two events are independent, we can use my third rule of probability about joint distributions, $P(X = A, Y = B) = P(X = A)$ $P(Y = B)$, and substitute this in

$$P(X = A | Y = B) = \frac{P(X = A, Y = B)}{P(Y = B)} = \frac{P(X = A)P(Y = B)}{P(Y = B)}$$

Since we have the probability of X in the top and bottom of the last formula, these will cancel out and we will get $P(X = A | Y = B) = P(X = A)$. So, in fact, when X and Y are independent, the conditional probability of X given Y is just equal to the probability of X. As we predicted, when X and Y are independent events, what happens with Y doesn't affect X.

RULES OF PROBABILITY

1. Probability of A or B is $P(X = A) + P(X = B)$, as long as A and B can't both happen.
2. Sum of probabilities has to be 1. $\sum_{A,B,C,...} P(X = A) = 1$.
3. Probability of A and B (aka. "joint" probability) $P(X = A, Y = B) = P(X = A)P(Y = B)$, if X and Y are independent.
4. Probability of A given B (aka. "conditional" probability) $P(X = A | Y = B) = P(X = A, Y = B) / P(Y = B)$.

Given these rules of probability, it's now possible to understand Bayes' theorem:

$$P(X = A | Y = B) = \frac{P(X = A)}{P(Y = B)} P(Y = B | X = A)$$

Bayes' theorem is a key result that relates conditional probabilities even if events are dependent. We'll use it over and over again in deriving statistical models. Another important idea that we'll see in this book is the expectation, which we denote as $E[X]$ or $\langle X \rangle$. You can think of this as our "best guess" about a random variable, where you sum over all the possible values that the random variable can have, weighting each by their probability.

$$E[X] = P(X = A)A + P(X = B)B + P(X = C)C + \cdots$$

An important thing to notice about the expectation is that it is a linear operator, so that $E[X + Y] = E[X] + E[Y]$ if X and Y are two random variables. Using the expectation operator, it's possible to define the variance as $V[X] = E[(X - E[X])^2]$.

NOTATION AND ABUSE OF NOTATION

I will do my best to use the simplest possible notations in this book, but in my efforts to do so, I will have to leave out some details and write things that aren't exact (or "abuse" the notation). For example, you probably already noticed that I gave some formulas for probability distributions as $p(x)$, and now I'm using a new notation $P(X = A)$. In the formula $P(X = A)$, I'm writing the probability that the random variable X takes the value A. Usually, I won't bother with this and I'll just write $P(X)$ or $p(X)$, and you'll have to infer from context whether I'm talking about the random variable or the *value* of the random variable. You'll see that this is easier than it sounds. For example,

$$E[X] = P(X = A)A + P(X = B)B + P(X = C)C + \cdots$$

$$E[X] = \sum_{i=1}^{n} P(X_i)X_i$$

$$E[X] = \sum_{X} p(X)X$$

All mean the same thing, with varying degrees of detail. I will usually choose the simplest formula I can, unless, for example, it's important to emphasize that there are n observations.

I'll be especially loose with my use of the notation for conditional probabilities. I'll start writing probability distributions as $p(X|\text{parameters})$ or null distributions as $p(X|H0 \text{ is true})$ implying that the parameters or hypotheses are actually observations of random variables. You might start thinking that it's a bit strange to write models with parameters as random variables and that this assumption might cast philosophical doubt on the whole idea of modeling as we discussed earlier. In practice, however, it's easier not to think about it—we're using statistical models because they describe our data. We don't really need to worry about exactly what the notation means, as long as we're happy with the predictions and data analysis algorithms that we get.

2.4 HYPOTHESIS TESTING: WHAT YOU PROBABLY ALREADY KNOW ABOUT STATISTICS

So what about *P*-values? What I actually took away from my introductory statistics courses was that somehow if an experiment was done properly, I could plug the numbers into a formula and it would spit out a very small *P*-value, which meant the experiment "worked." If the *P*-value wasn't that small, then the experiment didn't work—that was bad, because it meant that you probably had to do everything again. Although it's easy to understand why students get this impression, this is *completely wrong*.

P-values are the result of a so-called "statistical hypothesis test." For example, the *t*-test is specifically a test of whether the mean of two lists of numbers are different. In "statistical hypothesis testing" (the technical term for performing *t*-tests and other tests), the hypothesis that you are always testing is the so-called "null hypothesis." It's very important to realize that the "null hypothesis" that you are testing statistically is not usually the same as your actual (scientific) hypothesis. In fact, it's often exactly the opposite of the scientific hypothesis. In the case of the *t*-test, the null hypothesis is that the two lists of numbers truly have the same mean. Usually, the scientific hypothesis is that the two lists of numbers (e.g., describing mutant and wildtype) are different. Comparing lists of numbers is by far the most common use of statistical test in molecular biology, so we'll focus on those kinds of tests for the rest of this chapter. We'll discuss other hypothesis tests as they arise throughout the book.

Traditionally, hypothesis tests (such as the *t*-test) are generally set up in the following way. Define a "test statistic" (some function of the data) whose distribution is *known* under *the null hypothesis*, and look up the observed value of this test statistic in this so-called "null distribution" to test the probability that the data were drawn from the null hypothesis. Since finding a function of the data whose distribution was known (and possible to calculate by hand) was very unusual, inventing these tests was a great achievement. Statistical tests were therefore usually named after the distribution of the test statistic under the null hypothesis (the null distribution) and/or after the statistician who discovered the test statistic (and sometimes a combination of the two). So the null distribution for the (Student's) t-test is the Student's t-distribution because the test was proposed by a guy who called himself "Student" as a pseudonym.

Using the "rules of probability," we can write all this a bit more formally:

- $H0$ is the null hypothesis, which can be true or not.

- The observations (or data) are X_1, X_2, ..., X_N, which we will write as a vector X.

- t is a test statistic, $t = f(X)$, where f represents some defined function.

- The null distribution is therefore $P(t|H0$ is true).

- t^* is the observed value of the test statistic (for the set of data we have).

- The P-value is $P(t$ is "as or more extreme" than $t^*|H0$ is true), or $P(t \geq t^*|H0)$ in short.

Given these definitions, we also note that the distribution of P-values is also known under the null hypothesis:

- $P(P\text{-value} < p|H0$ is true) $= p$

In other words, under the null hypothesis, the P-value is a random variable that is uniformly distributed between 0 and 1. This very useful property of P-values will come up in a variety of settings. One example is "Fisher's method" for combining P-values. Given several different tests with P-values p_1, p_2, ..., p_n, you can combine them into a single P-value. Fisher figured out that the test statistic $t = f(p_1, p_2, ..., p_n) = -2 \sum_{i=1}^{i=n} \ln(p_i)$ has a known distribution if p_1, p_2, ..., p_n are uniformly distributed $\{0, 1\}$ and i.i.d. (It turns out to be approximately chi-squared distributed with $df = 2n$.) This type of test (a test on the P-values of other tests) is also called "meta-analysis" because you can combine the results of many analyses this way.

To illustrate the idea of hypothesis testing, let's take a real example based on gene expression data from ImmGen (Heng et al. 2008). Stem cells are expected to be rapidly dividing, so we might expect them to express genes involved in DNA replication, like Cdc6. Other cells, since they are differentiated and not duplicating their DNA, might be expected to show no (or low) expression of DNA replication genes like Cdc6 (Figure 2.4, top left). The *null hypothesis* is that stem cells and other cells show the *same* average Cdc6 expression level. Notice that this null hypothesis was actually the opposite of our biological hypothesis. We can do a *t*-test (on the expression levels from stem cells vs. the expression levels from other cells) and calculate a *P*-value to test the hypothesis. In this case, the *t*-statistic is −5.2979. If we look-up the probability of observing that *t*-statistic value or more in the null distribution (the known distribution of the *t*-statistic given the null distribution is true, which turns out to be a *T*-distribution), we get a *P*-value of 0.0001317. This means that the chances of stem cells actually having the same average expression levels are very small, which is great! This means that our data reject the null hypothesis, if the assumptions of the test are true.

And this brings up the caveat to all *P*-value calculations—these *P*-values are only accurate *as long as the assumptions of the test are valid*. Each statistical test has its own assumptions, and this is just something that you always have to worry about when doing any kind of data analysis. With enough data, we can usually reject the null hypothesis because we will have enough statistical power to identify small deviations between the real data and the assumptions of our test. It's therefore important to remember that in addition to the *P*-value, it's always important to consider the size of the effect that you'd identified: If you have millions of datapoints, you might obtain a tiny *P*-value for a difference in mean expression of only a few percent. We'll revisit this when we test for nucleotide content differences on chromosomes in Chapter 4.

The *t*-test assumes that the data are normally distributed (Figure 2.4, right panel), but turns out to be reasonably robust to violations of that assumption—the null distribution is still right when the data are only approximately normally distributed. The *t*-test tests for differences in the mean. Consider the data for CD4 antigen gene expression (also from ImmGen) in Figure 2.4. Clearly, the CD4 expression levels in T cells are more likely to be high than in the other cells, but the distribution is strongly bimodal—clearly not Gaussian (compare to the theoretical distribution in Figure 2.4). Although this probably does violate the assumptions of the *t*-test, the more important issue is that only testing for a difference in the mean expression level is probably not the best way to detect the difference.

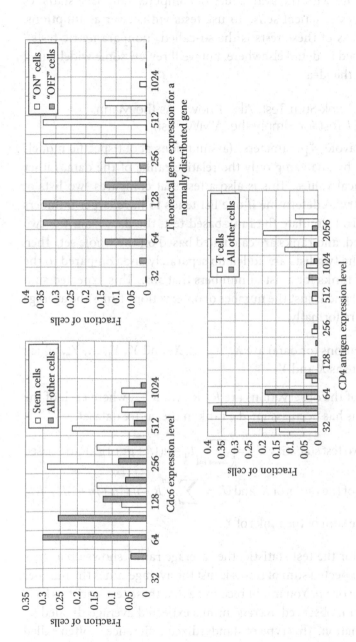

FIGURE 2.4 Testing hypotheses in practice with gene expression data from ImmGen. In each plot, the fraction of cells is plotted on the vertical axis as a function of the gene expression level on the horizontal axis. Real (Cdc6 and CD4) expression patterns (top left) differ from what is expected under a Gaussian model (top right). In the case of Cdc6, there is still a clear difference in the means, so a *t*-test would be just fine. In the case of CD4 (bottom left), there is a bimodal distribution, and a nonparametric test would probably work better.

2.5 TESTS WITH FEWER ASSUMPTIONS

In this context, and with the availability of computers and free statistics packages, it makes practical sense to use tests with fewer assumptions. An important class of these tests is the so-called "nonparametric tests." These are described in detail elsewhere, but we'll review some widely used examples to give the idea.

2.5.1 Wilcoxon Rank-Sum Test, Also Known As the Mann–Whitney U Test (or Simply the WMW Test)

The WMW test avoids "parameters" (assumptions about specific models or distributions) by analyzing only the relative ranks of the data, rather than the numerical values. This is also a test that compares two lists of observations to try to determine if one list tends to have larger numbers than the other. To formulate the rank-based test, the data from the two lists are combined, and ranks are calculated based on the whole set. Then the ranks from the two lists are added up separately and compared to the expected sum of ranks for a list of numbers that size. This expected sum can be calculated and is just the number of observations times the average rank. A little more formally,

- The observations (or data) are $X_1, X_2, \ldots, X_n$ and $Y_1, Y_2, \ldots, Y_m$, which we will write as X and Y.

- The ranks of the observations are $R_1, R_2, \ldots, R_{n+m}$, where each of the observations has been assigned a rank in the entire dataset.

- There are two test statistics: $U_X = \left(\sum_{i=1}^{i=n} R_i - n((n+m+1)/2) \right)/\sigma_U$ based on the sum of the ranks of X, and $U_Y = \left(\sum_{i=1}^{i=m} R_i - m((n+m+1)/2) \right)/\sigma_U$ based on the sum of the ranks of Y.

In the formulas for the test statistic, the "average rank" shows up as $(n + m + 1)/2$, so the expected sum of ranks is just the average times the number of observations (n or m). You might recognize that these test statistics have the following form: observed average minus expected average divided by the standard deviation. This type of standardized difference is often called a "Z-score." Amazingly, the formula for the standard deviation of the ranks also has a reasonably simple form, $\sigma_U = \sqrt{(mn/12)(m+n+1)}$. However, it's important to note that this formula assumes that there are no *tied ranks* in the data. This means that you never saw the same number twice. In practice, you'll be using a statistics package to calculate these tests, so just make sure that the software is handling the tied ranks properly.

Under the null hypothesis (that all the observations are drawn from a single distribution), these test statistics turn out to be (approximately) Gaussian distributed, with mean and standard deviation (0, 1). The *P*-value for the WMW test is the one associated with the *U*-statistic that is more extreme. Applying the WMW test to the examples mentioned earlier, we get $P = 0.00008$ for the Cdc6 data for Stem Cells, and $P = 0.00007$ for the CD4 data.

2.5.2 Kolmogorov–Smirnov Test (KS-Test)

The KS-test is another example of a popular, rank-based test. Once again, the KS-test compares two lists of numbers under the null hypothesis that they are actually drawn from the same pool, but this time it uses a test statistic based on the "cumulative distributions" of the observations. The cumulative distribution is the sum of the probability distribution up to a certain point. The KS-test uses as the test statistic the maximum difference between the cumulative distributions. Figure 2.5 illustrates the cumulative distribution and the KS-test statistic, usually referred to as *D*. Surprisingly, the distribution of this test statistic can be computed (approximately) under the null hypothesis, *regardless of the distribution of the data*. I reproduce the formula here only for aesthetic reasons—to show that there is no reason to expect that it should be simple:

$$P(D < D^* | H0) \approx 2 \sum_{j=1}^{\infty} (-1)^{j-1} \exp\left[-2j^2 D^{*2} \left(0.12 + \frac{0.11}{\sqrt{nm/(n+m)}} + \sqrt{\frac{nm}{n+m}} \right)^2 \right]$$

where
exp[*x*] is the exponential function
D^* is the observed value of the test statistic
n and *m* are the sizes of the two samples, as defined in the WMW section

Figure 2.5 illustrates why the KS-test would be expected to work reasonably well, even on a strongly bimodal dataset like the CD4 data. Like the WMW test, you will usually use a statistics package to perform this test, and be sure that if there are any tied-ranks in your data, the statistics software is handling them correctly. In the KS-test, it's particularly tricky to

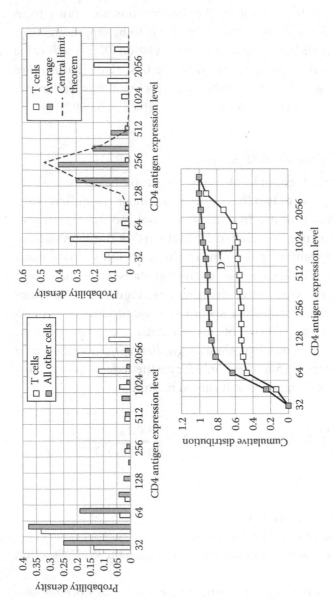

FIGURE 2.5 The KS-test and the central limit theorem apply even when the data are not distributed according to a standard distribution. T cells show a different distribution of expression levels for CD4 than other cells (top right) even though neither distribution looks Gaussian. The KS-test measures the difference between the cumulative distributions (D) between two samples (bottom left). Convergence to the central limit theorem is illustrated in the top left panel. Averages (gray bars, 20 samples of 20 datapoints, randomly chosen from the T-cell data, unfilled bars) show approximately Gaussian behavior (dashed line), even though the real data is strongly non-Gaussian.

handle the ties correctly. Once again the KS-test tells us that both these datasets are highly significant, with the Cdc6 data having $D = 0.6832$, $P = 1.038e{-}05$, while the CD4 data having $D = 0.3946$, $P = 5.925e{-}06$. Notice that the Cdc6 data have a larger D, but are not as significant (larger P-value). This is because there are more T cells than stem cells in the data-set. Here, you can see the nontrivial dependence on the size of the two sets in the formula for the KS-test P-value.

2.6 CENTRAL LIMIT THEOREM

The central limit theorem is one of the least understood and most powerful results in statistics. It explains, at some deep level, the connection between dice games and the shapes of Fisher's iris petals. The central limit theorem can also help you devise statistical tests (or interpret data) where the underlying observations are drawn from unknown or badly behaved distributions.

The key to understanding the central limit theorem is that it is about the distribution of means (averages), not the observations themselves. The amazing result is that the distribution of the means is known to be Gaussian, regardless of the distribution of underlying data, as long as they have a finite mean and variance. A little more formally,

- The observations (or sample) are $X_1, X_2, \ldots, X_N$.

- $A(X) = (1/N) \sum_{i=1}^{i=N} X_i$, is just the ordinary average, which will give a different answer each time N datapoints are sampled from the pool.

- If N is large enough, the distribution of $A(X)$ will be a Gaussian, $N(\mu, \sigma^2)$, where $\mu = E[X]$, the true expectation of X, and variance $\sigma^2 = V[X]/N$, the true variance of X divided by the sample size.

This means that for samples of badly behaved (non-Gaussian) data, like the CD4 data shown in the example, the distribution of averages will still be approximately Gaussian.

2.7 EXACT TESTS AND GENE SET ENRICHMENT ANALYSIS

Another important class of hypothesis tests that can be used to handle data with nonstandard distributions is related to Fisher's exact test. These types of tests can be applied to continuous data by first discretizing the data. For example, we can classify the CD4 data into "high" expression

TABLE 2.1   Numbers of Cells with High or Low
CD4 Expression Levels

	"High" (>256)	"Low" (<256)
T-cells	24	27
Other cells	17	146

cells and "low" expression cells by choosing a cutoff. For example, if we say cells with expression above 256 are "high" and cells with expression below 256 are "low" expression cells, we can form a two-by-two table (Table 2.1).

Note that there is a difference between T cells and the other cells: T cells are about 50% "high" expression, and all the other cells are less than 20% "high" expression. We now need to calculate the probability of having observed that big a difference in fraction (or ratio) under the null hypothesis that the two groups really have the same fraction (or ratio). A famous test for a table like this is Pearson's chi-squared test, which computes a test statistic based on the observed and expected numbers in each cell. In Pearson's test, however, the numbers in each cell have to be "large enough" so that the approximation for the null distribution starts to work. Traditionally, Pearson's test was used because it was easy to compute by hand, whereas Fisher's exact test was difficult (or impossible) to compute by hand. However, since we now use computers to do these tests, we can always use Fisher's exact test on this kind of data, which makes no assumptions about the numbers in each cell.

Unlike traditional hypothesis tests, exact tests don't really have a "test statistic" function that is computed based on the data. Instead, Fisher's test assigns a probability to every possible configuration of numbers in the table under the null hypothesis, that the numbers in each row and column were randomly sampled from the total pool of observations (using a formula that I won't reproduce here). To compute the P-value, one adds up all the configurations with probability smaller (more unlikely) than the observed configuration. Thus, the P-value is the probability of observing a configuration as unlikely (or more unlikely) under the null hypothesis. Computing this in practice requires a reasonably clever way to sum up the very large number of configurations, and modern statistics packages might do this in various ways.

A similar exact test is used very often in "Gene Set Enrichment Analysis" (Subramanian et al. 2005). In this context, one has identified a list of genes in a molecular biology experiment (a gene set) and wants to test whether the list of genes is random or whether the experiment has identified genes with specific biology associated ("enriched"). Initially, gene lists usually came from clustering of gene expression data, often from microarrays. Gene Set Enrichment Analysis was used to show enrichment of specific biological function for genes with similar expression patterns (with similarity measured in a high-dimensional space). We will return to this type of data analysis in Chapter 5.

Nowadays, Gene Set Enrichment Analysis is used on gene lists that arise from all types of data analysis. For example, we recently used the test in a bioinformatics project in my lab. We developed a new method to predict protein kinase-substrates based on amino acid sequences (Lai et al. 2012) and wanted to know whether the set of predicted substrates contained more previously known substrates than expected by chance. We predicted 46 Mec1 substrates, and of these 7 were already known to be substrates. Considering there were only 30 known substrates in the databases, and we analyzed 4219 proteins in total, we thought we were doing pretty well: Our list had 15% known substrates, while fewer than 1% of proteins are known to be substrates. A little more formally

- We have a sample, X_1, X_2,..., X_n, where each observation can either have a certain property or not, which we denote as "positives" ($X_i = 1$) or "negatives" ($X_i = 0$).

- The probability of observing the number positives $k = \sum_{i=1}^{n} X_i$ if X was a random sample from some finite pool, Y_1, Y_2,..., Y_m, with $l = \sum_{i=1}^{m} Y_i$ positives total is given by the hypergeometric distribution

$$P_{HYP}(k \mid n,l,m) = \frac{\binom{l}{k}\binom{m-l}{n-k}}{\binom{m}{n}}$$

- The P-value is $P(\geq k|H0) = \sum_{i=k}^{i=\min(n,l)} P_{HYP}(i|n,l,m) = 1 - \sum_{i=0}^{i=k-1} P_{HYP}(i|n,l,m)$, where the last equality $P(\geq k|H0) = 1 - P(<k|H0)$ is just used to make the calculations easier.

In practice, this type of exact test can usually only be calculated using a reasonably clever statistics package, because to calculate

$$\binom{m}{n} = \frac{m!}{n!(m-n)!}$$

in many examples requires computing very large factorials (e.g., 4219! = $1 \times 2 \times 3 \times \cdots \times 4218 \times 4219$ is too large to be stored as a standard number on a computer).

In the example, we want to calculate the probability of getting 7 or more known substrates in 46 predictions, when there are 30 known substrates in the database of 4219 proteins. This works out to

$$P_{HYP}(\geq 7|30, 46, 4219) \approx 2 \times 10^{-8}$$

Needless to say, this is a very small number, supporting the idea that the list of Mec1 predictions was very unlikely to overlap this much with the known substrates by chance. Thus, the gene set was "enriched" for previously known substrates.

2.8 PERMUTATION TESTS

What if a situation arises where you have a hypothesis that isn't summarized by any traditional hypothesis test? For example, say you wanted to test if the mean is bigger *and* the standard deviation is smaller. You might want to use the test statistic $a(X) - sd(X)$ (where I used $a()$ to denote the mean and $sd()$ for the standard deviation). Or maybe you think (for some reason) that the ratio of the maximum to the average is what's important in some conditions. You might use the test statistic $max(X)/a(X)$. You might try to figure out the distribution of your test statistic under the null hypothesis, assuming your data were Gaussian.

However, another way to do it is to use your data to compute the null distribution using a permutation test. These tests are particularly powerful when you are comparing two samples (such as T cells vs. other cells, or predicted "positives" vs. "negatives"). In all of these cases, the null hypothesis is that the two samples are actually *not* different, i.e., they are drawn from the same distribution. This means that it's possible to construct an estimate of that distribution (the null distribution) by putting the observations from the two sets into one pool, and then drawing randomly the observations in the two samples. In the case of the CD4 measurements in

T cells versus other cells, we would mix all the T-cell measurements with the other cell measurements, and then randomly draw a sample of the same size as the T-cell sample from the combined set of measurements. Using this random fake T-cell sample, we can go ahead and compute the test statistic, just as we did on the real T-cell data. If we do this over and over again, we will obtain many values for the test statistic (Figure 2.6, left panel). The distribution of the test statistic in these random samples is exactly the null distribution to compare our test statistic to: These are the values of the test statistic that we *would* have found if our T-cell sample was truly randomly drawn from the entire set of measurements. If we sample enough times, eventually we will randomly pull the exact set of values that are the real T-cell data and get the exact value of the test statistic we observed. We will also, on occasion, observe test statistics that exceed the value we observed in the real data. Our estimate of the P-value is nothing more than the fraction of random samples where this happens (Figure 2.6, right panel). This approach is sometimes referred to as "permuting the labels" because in some sense we are forgetting the "label" (T-cell or not) when we create these random samples.

The great power of permutation tests is that you can get the null distribution for any test statistic (or any function of the data) that you make up. Permutation tests also have several important drawbacks: Permutation tests involve large numbers of random samples from the data—this is only

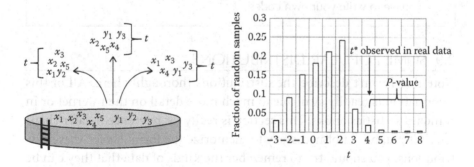

FIGURE 2.6 Illustration of how to obtain a P-value from permutation test. It indicates a test statistic, which is just a function of a sample of numbers from the pool. By choosing random samples from the pool according to the null hypothesis and computing the test statistic each time, you can calculate the null distribution of the test statistic. To calculate the P-value, you sum up the area under the null distribution that is more extreme than the value you observed for your real data.

possible with a computer, and if the number of samples needed is large or complicated, it could take a very long time to get there. Probably, more of an issue in practice is that for an arbitrary test statistic, the permutation test needs to be "set up" by using some kind of programmable statistics package. Finally, in order to implement a permutation test, you have to really understand your null hypothesis. This can become especially tricky if your data are correlated or have complicated heterogeneity or structure.

KEY STATISTICAL TESTS FOR COMPARING TWO LISTS OF NUMBERS

- *t-Test*: Finds differences in means between two lists of numbers (think mutant vs. wt) when data are approximately Gaussian. Works for small sample sizes.
- *Wilcoxon Rank-Sum/WMW test*: Nonparametric version of the *t*-test. Works on any data distribution, but less power than a *t*-test, especially for small sample sizes.
- *KS-test*: Nonparametric test for differences in distributions of two lists of numbers. Again, less power for small sample sizes. Most powerful when there is a difference in means, but also pretty good in general.
- *Fisher's exact test*: Tests for differences in ratios or fractions (data in a 2 × 2 table). Works on any sample size. Can be used on continuous data if a cutoff is chosen to define classes.
- *Permutation test*: Roll your own test statistic and estimate the null distribution by resampling your data. Since it's custom, you usually have to write your own codes.

2.9 SOME POPULAR DISTRIBUTIONS

Note that I won't describe the distributions thoroughly here. All of this information is readily available in much more detail on the Internet or in standard statistics books. This section is really for reference in later chapters. Although it's not necessary to memorize the formulas for these distributions, you should try to remember the kinds of data that they can be used to model.

2.9.1 The Uniform Distribution

This is the simplest of all distributions. It has a finite maximum and minimum, and assigns the same probability to every number in the interval. So the probability of observing X is just 1/(maximum − minimum). Some people even use the word "random" to mean uniform distributed.

2.9.2 The T-Distribution

The T-distribution is a so-called "heavy-tailed" approximation to a Gaussian distribution (with mean = 0 and standard deviation = 1). This means that it looks a lot like the normal distribution, except that the probability in the tails of the distribution (the parts of the distribution far away from the peak or mode) doesn't decay as fast as in the Gaussian. The similarity of the T-distribution to the Gaussian (with mean = 0 and standard deviation = 1) is controlled by the only parameter of the distribution, which is called the "degrees of freedom" or df for short. When $df = 1$ or 2, the tails of the T-distribution are quite a bit heavier than the Gaussian, so the probability of getting observations far away from the mean is large. When df is around 30, the T-distribution starts to look a lot like a Gaussian. Of course, only when $df = $ infinity is the T-distribution *exactly* equal to the Gaussian distribution.

The formula for the T-distribution is a bit complicated, so I won't reproduce it here.

2.9.3 The Exponential Distribution

$$P(X|\lambda) = \lambda e^{-\lambda X} \quad \lambda > 0, X > 0$$

A heavy-tailed distribution defined for positive real numbers. The exponential distribution is the classic model for waiting times between random events.

2.9.4 The Chi-Squared Distribution

This is another distribution that models real numbers greater than zero. It also has only one parameter, known as the "degrees of freedom" or df. Mathematically, it corresponds to the square of the Gaussian distribution, but I won't reproduce the formula here because it's a bit complicated. The chi-squared distribution is rarely used to model actual data, but it is an important distribution because many widely used statistical tests have chi-squared distributions as their null distributions (e.g., Pearson's chi-square test for 2×2 tables).

2.9.5 The Poisson Distribution

$$P(X|\lambda) = \frac{e^{-\lambda}\lambda^X}{X!} \quad \lambda > 0, X \in \{0,1,2,...\}$$

The Poisson is the most widely used distribution defined on natural numbers (0, 1, 2, ...). Another distribution on natural numbers is the geometric distribution, which is the discrete analog of the exponential distribution.

2.9.6 The Bernoulli Distribution

$$P(X|f) = f^X (1-f)^{1-X} \quad f > 0, X \in \{0,1\}$$

This is the classic "heads" and "tails" distribution for a single toss of a coin. In my notation, we arbitrarily assign heads to be 1 and tails to be 0.

2.9.7 The Binomial Distribution

This is the Bernoulli distribution when you've made n observations:

$$P(X|f,n) = \frac{n!}{\left(\sum_{i=1}^{n} X_i\right)! \left(n - \sum_{i=1}^{n} X_i\right)!} f^{\sum_{i=1}^{n} X_i} (1-f)^{\left(n - \sum_{i=1}^{n} X_i\right)}$$

$$f > 0, X_i \in \{0,1\}$$

In this formula, I've indexed the n observations using i, and I hope it's clear that $\sum_{i=1}^{n} X_i$ and $n - \sum_{i=1}^{n} X_i$ are the numbers of "heads" (or 1s) and "tails" (or 0s) respectively.

EXERCISES

1. Derive Bayes' theorem.

2. Show that Bayes' theorem is true even if the two events are independent.

3. Show that the geometric distribution is correctly normalized (i.e., the sum of all possible observations is 1). (*Hint*: Use the classical result that $\sum_{n=0}^{\infty} a^n = 1/(1-a)$.)

4. It's not quite true that $-\infty$ to ∞ is the only range for which the Gaussian integral can be evaluated. There is one other range. What is it, and what's the value of the integral? (*Hint*: No calculations required—just think about the shape of the Gaussian distribution.)

5. Explain why $n + m + 1$ divided by 2 is the expected average rank in the WMW test. (*Hint*: Use the classical result that $\sum_{j=1}^{N} j = (N(N+1))/2$.)

6. Use the central limit theorem to argue that the null distribution of the WMW test statistic should be Gaussian.

7. Consider the following two lists of numbers. The first list is 1, 2, then ten 3s, then 4 and 5. The second list is the same, except there are one hundred 3s (instead of 10). Draw the cumulative distributions for these two lists, and explain why this might lead to a problem for the KS-test if ties were not handled correctly.

8. In the P-values for Gene Set Enrichment Analysis, why does the first sum go up only until $\min(n, l)$ and the second one go up to $k - 1$?

9. Look up (or plot) the shapes of the Poisson and Geometric distribution. Speculate as to why the Poisson is used more often.

REFERENCES AND FURTHER READING

Fisher RA. (1936). The use of multiple measurements in taxonomic problems. *Ann. Eugen.* 7(2):179–188.

Heng TS, Painter MW, Immunological Genome Project Consortium. (October 2008). The Immunological Genome Project: Networks of gene expression in immune cells. *Nat. Immunol.* 9(10):1091–1094.

Lai AC, Nguyen Ba AN, Moses AM. (April 1, 2012). Predicting kinase substrates using conservation of local motif density. *Bioinformatics* 28(7):962–969.

McAdams HH, Arkin A. (February 4, 1997). Stochastic mechanisms in gene expression. *Proc. Natl. Acad. Sci. USA.* 94(3):814–819.

Shahrezaei V, Swain PS. (November 11, 2008). Analytical distributions for stochastic gene expression. *Proc. Natl. Acad. Sci. USA.* 105(45):17256–17261.

Shalek AK, Satija R, Shuga J, Trombetta JJ, Gennert D, Lu D, Chen P et al. (June 19, 2014). Single-cell RNA-seq reveals dynamic paracrine control of cellular variation. *Nature* 510(7505):363–369.

Subramanian A, Tamayo P, Mootha VK, Mukherjee S, Ebert BL, Gillette MA, Paulovich A et al. (October 25, 2005). Gene set enrichment analysis: A knowledge-based approach for interpreting genome-wide expression profiles. *Proc. Natl. Acad. Sci. USA.* 102(43):15545–15550.

CHAPTER 3

Multiple Testing

3.1 THE BONFERRONI CORRECTION AND GENE SET ENRICHMENT ANALYSIS

Given the ease with which data can be collected and statistical tests can be performed with computational statistics packages, the problem in contemporary molecular biology is that we often have too many statistical tests that we might want to do. For example, after a high-throughput screen we are typically in the situation where we have obtained a list of genes from the experiment, and we have no idea what those genes have to do with our experiment. We can try to come up with a hypothesis or two, but it's often more convenient to do systematic gene set enrichment analysis. We test for all possible enrichments of anything we can get our hands on: gene ontology (GO) categories, Munich Information Center for Protein Sequences (MIPS) functions, Kyoto Encyclopedia of Genes and Genomes (KEGG) pathways, ChIP-seq compendia, phosphoproteomics databases, ... you get the picture. In bioinformatics jargon, these types of functional assignments to genes are known as "annotations." Table 3.1 shows some examples of functional annotations.

Of course, this seems like a good idea, and we can often discover interesting biology this way. However, let's think about the statistical situation here. As discussed in Chapter 2, we calculate an exact P-value using the hypergeometric distribution. This P-value is the probability of getting a gene list as enriched or more as the one we got under the null hypothesis that our gene list was randomly drawn from the pool of genes. Let's say we want the significance threshold to be $P < 0.05$. This threshold means

TABLE 3.1 Some Common Annotations of Gene Function

Gene ontology (GO)	Assigns biological process, cellular compartments, and molecular functions based on a variety of evidence. Annotations or "GO terms" are organized into a hierarchy, where most general terms at the top, and most specific terms are at the bottom. Terms at the bottom can be thought to "inherit" all the terms above them. For example, "MAP kinase phosphorylation," would be below "protein phosphorylation" in hierarchy.
Munich Information Center for Protein Sequences (MIPS)	Comprehensive, consistent automatic annotation based on sequences and curation of protein interactions.
Kyoto Encyclopedia of Genes and Genomes (KEGG)	Most widely used for assignments of genes into biological pathways. Each pathway is either metabolism, genetic information processing, environmental information processing, organismal systems, human diseases, or drug development. Within these broad categories, pathways such as "amino acid metabolism" or "replication and repair" are constructed from gene products.
Reactome	Human genes are assigned into curated pathways. These are organized hierarchically into large pathway groups like "cell cycle." Specific reactions like "Orc3 associates with Orc2" falls under cell cycle.

that we want there to be a 1 in 20 chance of seeing a gene list this enriched (or more) under the null hypothesis (where there was truly no enrichment). Consider what happens if we now have 20 different annotations to test: although each one of them *alone* has a 1/20 chance of getting a P-value of $P < 0.05$, because we did 20 tests, *overall* we expect one of them to have $P < 0.05$ under the null hypothesis (when there is no enrichment!). More generally, for any P-value threshold, α, if we do m tests, we expect about m times α tests to pass it under the null hypothesis. Another way to think about this is to remember that the distribution of P-values under the null hypothesis is uniform between 0 and 1. 0.05 covers 1/20 of this distance. This means that if we randomly draw m P-values from the null distribution, we expect about $m/20$ will fall below 0.05.

You can see immediately that in the context of many tests (or multiple testing) there will be a problem with our conventional use of the P-value. The simplest solution to this problem is the so-called Bonferroni correction to the P-value. The idea here is to use what we know about the distribution of P-values under the null hypothesis to correct for the number of tests we've done.

Since we know the probability distribution of the P-values under the null hypothesis is uniform, we simply *rescale* the region that corresponds to a "significant" P-value to be 0.05 *taking into account* the fact that we have done m tests. The new P-value threshold is simply $0.05/m$ and now there will only be a 5% chance of observing a P-value that small under the null hypothesis. Alternatively, we can achieve the same effect by rescaling the P-values themselves or "correct" the P-values by simply multiplying by the number of tests: A P-value, P, becomes $m \times P$. To keep the P-values as probabilities, any P-value that gets bigger than one is simply set to be one. Formally, the correction can therefore be written as

$$P_{Bonferroni} = \min[1, m \times P]$$

where min[] indicates the minimum of whatever is in the square brackets. Now, no matter how many tests we do, we will only have a 5% chance of observing a P-value below 0.05 under the null hypothesis. The Bonferroni correction is widely used in practice because it is so simple to calculate. The only tricky part is knowing exactly what number of tests you really did (this is not always strictly equal to the number of tests you actually performed using your statistics software).

To illustrate just how tricky multiple testing can be, let's consider a realistic example. Let's say you were given a gene list with five mouse genes that came out of genetic screen. Not knowing anything else about the genes, you look at the GO annotations and notice that three of the genes are annotated with the GO term GO:0001525, angiogenesis. Since there are only 218 other mouse genes annotated with this term out of 24,215 mouse genes, this looks like a dramatic enrichment: 60% of your list is associated with angiogenesis, and less than 1% of the mouse genes are. Indeed, the exact test described in Chapter 2 for gene set enrichment seems to support this: $P_{HYP}(\geq 3 | 218, 5, 24,215) \approx 7 \times 10^{-6}$.

However, when we did this test we had no hypothesis, so we would have accepted enrichment of any GO annotations. And since there are $m = 16,735$ different GO annotations assigned to mouse genes, the Bonferroni correction tells us we should multiply the P-value by 16,735. I have

$$P_{Bonferroni} = \min[1, m \times P] = \min[1, 16,735 \times P_{HYP}(\geq 3 | 218, 5, 24,215)] \approx 0.12$$

This P-value does not pass a nominal 0.05 threshold after the Bonferroni correction. This means that if we choose random lists of five genes 20 times, we're likely to see a list of five genes that has an enrichment as strong as what we observed for the GO term GO:0001525, angiogenesis.

3.2 MULTIPLE TESTING IN DIFFERENTIAL EXPRESSION ANALYSIS

Another very well-studied example where multiple testing is important in modern molecular biology is trying to identify differentially expressed genes. Let's consider trying to find the genes that are differentially expressed between T cells and all other cells in the ImmGen data. In Chapter 2, we did this for a couple of genes, but, in fact, we have ~25,000 of genes whose expression has been measured in hundreds of cells. More formally,

- We want to find for which of m genes the expression level is different in the T cells.

- For each gene, j, we have a sample of n observations, X_{j1}, X_{j2}, ... and Y_{j1}, Y_{j2}, ... where X are the expression levels in the T cells and Y are the expression levels in other cells.

- The null hypothesis, $H0$, is that the expression level in the T cells (X) and other cells (Y) is actually the same.

- For each gene, we compute a test statistic, $t^* = f(X_j, Y_j)$ and compute the P-value, or $P_j(t \geq t^*|H0)$, where I've now indexed the P-value with j, to indicate that we have one P-value for each gene.

Under the null hypothesis, for any P-value threshold we choose α, we will expect to find m times α genes beyond that threshold under the null hypothesis. The Bonferroni correction for multiple testing says that we should choose that threshold to be α/m, so that we expect less than one gene to pass this threshold under the null hypothesis. Figure 3.1 shows the results of actually doing 24,922 t-tests on the ImmGen data.

First of all, notice that the y-axis of these graphs is in *log* scale, so the number of genes with $P < 0.05$ is very large. In fact, more than 50% of the 25,000 genes are expressed differently in the T cells compared to all other cells. Obviously, this is more than the 5% expected with $P < 0.05$ by chance. In fact, by looking at the distribution of P-values we can guess that probably around 300 genes (halfway between 100 and 1000 on a log scale) are expected in each bin—that's the number we see in the bins for P-values greater than about 0.3. Based on this, you'd guess that of the more than 14,000 genes with P-value < 0.05 (remember, log scale) there are

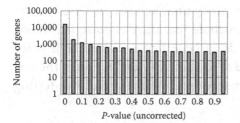 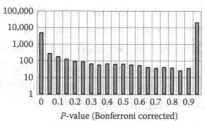

FIGURE 3.1   Identifying differentially expressed T-cell genes. The *y*-axis shows the number of tests with a *P*-value within the bin indicated for either uncorrected or raw *P*-values (left) or Bonferroni corrected *P*-values (right).

14,000 − 300 = 13,700 more than expected by chance. Alternatively, you could argue that you would have expected to see 5% of the 25,000 under the null hypothesis in each bin, or about 1,250. This suggests that there are 14,000 − 1,250 = 12,750 more than expected under the null hypothesis. So based on the uncorrected *P*-values, in either case we'd guess that more than 12,000 genes have different expression patterns in the T cells relative to the other cells.

Now let's consider the effect of the Bonferroni correction (Figure 3.1, right panel). You can see that the correction has shifted many *P*-values into the largest bin: These were all the *P*-values that were set to 1 because *m* times *P* was greater than 1. But more importantly, look at the 0.05 bin (furthest to the left). There are now 4500 genes with *P* < 0.05. Since the *P*-values are now corrected, we can say that these are the genes that have significantly different expression in T cells compared to all other cells. Yay! 4500 genes is a lot of genes. However, you might have noticed that this is a lot less than the more than 12,000 genes we estimated by looking at the uncorrected *P*-value distribution. Essentially, what's going on here is that the Bonferroni correction is giving us the 4500 *most* significantly different genes, but in fact it's being *way* too stringent. The Bonferroni correction was designed for situations where only one or two or a handful of tests would be significant—it really doesn't make much sense in the case where 1000s of tests are significant. It's not really reasonable to expect to do thousands of tests and not include *any* false rejections of the null hypothesis—when we require that *everything* above the threshold is truly significant (using the Bonferroni correction) we're going to miss 1000s of significant things that happen to fall below the threshold—in this case, the bulk of our signal. In this example, we still found plenty of significant genes, so we felt happy, but there are many situations, where based on the Bonferroni correction, we would conclude that there is no signal in our data, but in fact there are thousands of *bona fide* associations. It is for this reason that it's always advisable to look at this distribution of the raw P-values for at least a few examples before deciding on the multiple-testing correction procedure.

3.3 FALSE DISCOVERY RATE

From a probabilistic perspective, the interpretation of the Bonferroni correction is that we choose a new threshold for the P-value where we don't expect *anything* to pass by chance. In practice, however, with more than 25,000 GO terms used in gene set enrichment analysis and 25,000 genes measured in each experiment, there are situations where we just won't be able to get a small enough P-value to pass the Bonferroni corrected threshold.

In these cases, we might choose another threshold where we know that some of the tests that pass are actually false discoveries (expected under the null hypothesis). We might be willing to accept some of these mistakes, as long as the fraction of tests that are false discoveries is known (and small). Changing the P-values according to this strategy is known as controlling the false discovery rate (FDR) or performing an FDR correction.

The most famous and widely used FDR correction is the so-called Benjamini–Hochberg correction (Benjamini and Hochberg 1995). This correction is applied as follows:

1. Calculate the P-values for each of the tests that are performed.

2. Rank the tests by their test statistics so that the most significant is 1st, the second most significant is 2nd, etc.

3. Move down the ranked list, and reject the null hypothesis for the r most significant tests, where the P-value is less than $r\alpha/m$, where m is the number of tests and α is the FDR threshold.

In this case, α is no longer a P-value, but rather an estimate of the fraction of the tests that are falsely rejected, or $P(H0$ was true$|H0$ was rejected$)$. Notice the similarity between the formula to choose the FDR P-value threshold and the Bonferroni. For $r = 1$, there is only one test with a small P-value; the FDR is equivalent to the Bonferroni correction. However, if there are 100 tests with small P-values, now the threshold will be 100 times less stringent than the Bonferroni correction. However, with FDR, we accept that $100 \times \alpha$ of those tests will actually be false.

Applying the Benjamini–Hochberg (BH) procedure to the search for genes that are differentially expressed between T cells and other cells, at $\alpha = 0.05$, we have more than 13,000 genes passing the threshold. Since 5% of these are false discoveries, our estimate for the number of genes is actually closer to 12,000. Notice that this is very close to what we guessed by looking at the P-value distribution mentioned earlier, and more than 2.5× the estimate we got using the Bonferroni correction.

There are other interesting ways of controlling the FDR (Storey 2002) but the BH procedure is widely used because it is simple to calculate.

3.4 eQTLs: A VERY DIFFICULT MULTIPLE-TESTING PROBLEM

As a final example of how important multiple testing can be in practice, let's consider one of the most complicated statistical problems in modern molecular biology: the search for expression quantitative trait loci or eQTLs. The goal of this experiment is to systematically find associations between gene expression levels and genotypes. For example, in a classic eQTL experiment in yeast (Brem and Kruglyak 2005) the expression levels for 1000s of genes were measured in more than 100 segregants of a cross of a BY strain with an RM strain. They also genotyped the segregants at 1000s of markers covering nearly the whole genome. eQTLs are genetic loci (or markers, usually represented by single nucleotide polymorphsism [SNPs]) that are associated with gene expression levels. For example, there was a very strong eQTL for AMN1 expression just upstream of the AMN1 gene, so that strains with the RM allele had much higher gene expression levels than those with the BY allele (Figure 3.2, top left). This is exactly the sort of thing that we're looking for in an eQTL study, as it strongly suggests that there's an important genetic difference between the two strains that controls expression of the nearby gene.

As obvious from the histogram, there's a very strong statistical difference between these two expression patterns. Since the expression levels look reasonably Gaussian, we can go ahead and do a t-test, which gives a P-value around 10^{-30}, strong evidence that the genotype and expression level are not independent.

However, *AMN1* might not be the only gene whose expression is linked to this locus. Sure enough, *DSE1*'s expression level is also very strongly linked to the SNP near the *AMN1* locus, and as you can guess from the

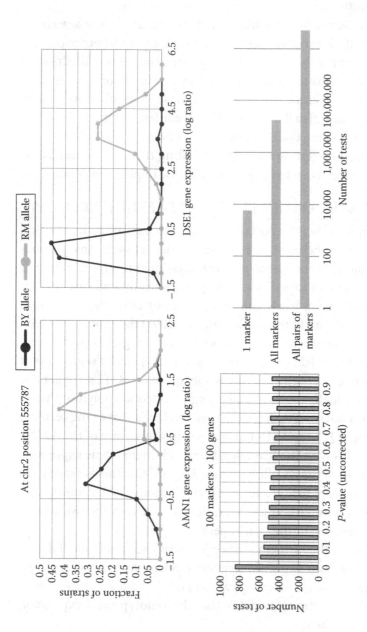

FIGURE 3.2 Multiple testing in a yeast eQTL experiment. The top two panels show the distribution of gene expression levels for *AMN1* (left) and *DSE1* (right) colored by the genotype at Chromosome 2, position 555787. The bottom panel's left panel shows the distribution of *P*-values for *t*-tests to identify eQTLs for 10,000 random gene expression level, marker combination. The bottom right panel shows the number of tests needed to discover eQTLs for either one position in the genome (1 marker), all positions in the genome (all markers) or all pairs of genotypes (all pairs of markers).

histogram, it's highly statistically significant (Figure 3.2, top right). And this is the great thing about the eQTL experiment: we've measured expression levels for all ~5800 yeast genes. So we can go ahead and test them all against the genotype at this locus, it's only 5800 tests: Clearly, these will still be significant even after a stringent multiple-testing correction. However, in general, we don't know which will be the important loci to look at: remember, we also have the genotypes at ~3000 loci. We need to try them all; now we're up to more than 17 million tests. The P-values for 10,000 of those WMW tests (randomly chosen from the 17 million) are shown in the bottom left panel of Figure 3.2. You can see that like the differentially expressed genes, there are already a large number of tests piling up in the bin for $P < 0.05$, so that some kind of FDR correction will be probably be appropriate here.

In general, however, geneticists don't believe that genotypes affect quantitative traits independently: typically, the expression level of a given gene is expected to be affected by multiple loci in the genome (e.g., genetic differences in transcription factor binding sites in the promoter, as well as genetic differences in the transcription factors that bind to them). What we would really like is to test for associations between combinations of genotypes and the expression levels. I hope it's clear that doing this naively even just for pairs of loci will yield more than 10 billion tests, not to mention the computer power need to actually compute all the tests (Figure 3.2, bottom left). Obviously, smarter methods to search for these types of effects are needed.

Finally, to add one more layer of complexity to the eQTL picture: So far, we've been considering only searching for differences in gene expression level that are associated with genotype. In fact, it's quite possible that the variability in gene expression levels or covariance between expression levels of one or more genes is actually under genetic control: This means that we could consider an even larger number of phenotypes, derived from the single-gene measurements. I hope that it's no surprise that development of statistical methods for eQTL analysis has been an area of intense research interest.

EXERCISES

1. In Chapter 2, I did a test on a gene list (the predicted substrates of Mec1). I did not correct for multiple testing. Why was this justified? What was the difference between that and the example gene list in this chapter?

2. I taught a graduate course where both professors had the same last initial: Prof. M and Prof. M. Were students in our class witness to something highly unlikely? (*Hint*: Assume that last initials are uniformly distributed and correct for multiple testing.) In fact, last names are not uniformly distributed. What effect is this expected to have on the probability?

3. A classic probability problem is the "birthday problem" where students in a class are surprised to learn that two of them either have the same birthday, or birthdays within a few days. Explain how the "surprise" in the birthday problem is really due to a failure to account for multiple testing.

4. Assume that you were trying to do an eQTL study using expression of all the genes in the ImmGen dataset on all the 1.5 million SNPs in the phase 3 HapMap. What would be the multiple testing correction?

REFERENCES AND FURTHER READING

Benjamini Y, Hochberg Y. (1995). Controlling the false discovery rate: A practical and powerful approach to multiple testing. *J. Royal Stat. Soc. Ser. B* 57:289–300.

Brem RB, Kruglyak L. (February 1, 2005). The landscape of genetic complexity across 5,700 gene expression traits in yeast. *Proc. Natl. Acad. Sci. USA* 102(5):1572–1577.

Storey J. (2002). A direct approach to false discovery rates. *J. Royal Stat. Soc. Ser. B* 64:479–498.

CHAPTER 4

Parameter Estimation and Multivariate Statistics

HAVING REVIEWED BASIC STATISTICS in the previous chapters, we now turn to the statistical background we need to understand the probability models that underlie simple statistical modeling techniques and some of the probabilistic machine learning methods. Although not all data analysis techniques and machine learning methods are based on statistical models, statistical models unify many of the basic machine learning techniques that we will encounter later in the book.

4.1 FITTING A MODEL TO DATA: OBJECTIVE FUNCTIONS AND PARAMETER ESTIMATION

Given a distribution or "model" for data, the next step is to "fit" the model to the data. Typical probability distributions will have unknown parameters (numbers that change the shape of the distribution). The technical term for the procedure of finding the values of the unknown parameters of a probability distribution from data is "estimation." During estimation, one seeks to find parameters that make the model "fit" the data the "best." If this all sounds a bit subjective, that's because it is. In order to proceed, we have to provide some kind of objective (mathematical) definition of what it means to fit data the best. The description of how well the model fits the data is called the "objective function." Typically, statisticians will try to find "estimators" for parameters that maximize (or minimize) an objective function. And statisticians will disagree about which estimators or objective functions are the best.

In the case of the (univariate) Gaussian distribution, the parameters are called the "mean" and the "standard deviation" often written as μ (mu) and σ (sigma). In using the Gaussian distribution as a model for some data, one seeks to find values of mu and sigma that fit the data. As we shall see, what we normally think of as the "average" turns out to be an "estimator" for the μ parameter of the Gaussian.

4.2 MAXIMUM LIKELIHOOD ESTIMATION

The most commonly used objective function is known as the "likelihood," and the most well-understood estimation procedures seek to find parameters that maximize the likelihood. These maximum likelihood estimates (once found) are often referred to as MLEs.

The likelihood is defined as a conditional probability: $P(data|model)$, the probability of the data give the model. Typically, the only part of the model that can change is the parameters, so the likelihood is often written as $P(X|\theta)$ where X is a data vector (or matrix), and θ is a vector containing all the parameters of the distribution(s). This notation makes explicit that the likelihood depends on both the data and the parameters in the model. Since, during parameter estimation, the data is assumed to be fixed, but the parameters can change, sometimes the likelihood is written as $L(\theta|X)$, but I find this terribly confusing, so I will not do that again.

More formally, let's start by saying we have some i.i.d., observations from a pool, say $X_1, X_2, \ldots, X_n$, which we will refer to as a vector X. We want to write down the likelihood, L, which is the conditional probability of the data given the model. In the case of independent observations, we can use the joint probability rule to write:

$$L = P(X|\theta) = P(X_1|\theta)P(X_2|\theta),\ldots,P(X_n|\theta) = \prod_{i=1}^{n} P(X_i|\theta)$$

Maximum likelihood estimation says: "choose the parameters so the data is most probable given the model" or find θ that maximizes L. In practice, this equation can be very complicated, and there are many different analytic and numerical techniques to solve it.

4.3 LIKELIHOOD FOR GAUSSIAN DATA

To make the likelihood specific, we have to choose the model. If we assume that each observation is described by a Gaussian distribution, we have two parameters—the mean (μ) and the standard deviation (σ).

$$L = \prod_{i=1}^{n} P(X_i|\theta) = \prod_{i=1}^{n} N(X_i|\mu,\sigma) = \prod_{i=1}^{n} \frac{1}{\sigma\sqrt{2\pi}} e^{-(X_i-\mu)^2/2\sigma^2}$$

Admittedly, the formula looks complicated. But in fact, this is a *very* simple likelihood function. I'll illustrate this likelihood function by directly computing it for an example (Table 4.1).

Notice that in this table I have chosen values for the parameters, μ and σ, which is necessary to calculate the likelihood. We will see momentarily that these parameters are *not* the maximum likelihood estimates (the parameters that maximize the likelihood), but rather just illustrative values. The likelihood (for these parameters) is simply the product of the five values in the right column of the table.

It should be clear that as the dataset gets larger, the likelihood tends to get smaller and smaller, but always is still greater than zero. Because the likelihood is a function of all the parameters, even in the simple case of the Gaussian, the likelihood is still a function of two parameters (plotted in Figure 4.1) and represents a surface in the parameter space. To make this figure, I simply calculated the likelihood (just as in Table 4.1) for a large number of pairs of mean and standard deviation parameters. The maximum likelihood can be read off (approximately) from this graph as the place where the likelihood surface has its peak. This is a totally reasonable way to calculate the likelihood if you have a model with one or two

TABLE 4.1   Calculating the Likelihood for Five Observations under a Gaussian Model

Observation (i)	Value (X_i)	$P(X_i|\theta) = N(X_i|\mu = 6.5, \sigma = 1.5)$
1	5.2	0.18269
2	9.1	0.059212
3	8.2	0.13993
4	7.3	0.23070
5	7.8	0.18269
$L = 0.000063798$		

Note: Each row of the table corresponds to a single observation (1–5).

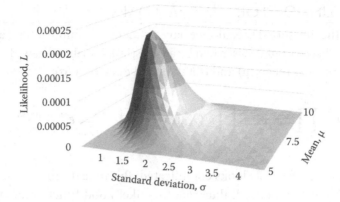

FIGURE 4.1 Numerical evaluation of the likelihood function. The surface shows the likelihood as a function of the two parameters (the mean and standard deviation).

parameters. You can see that the standard deviation parameter I chose for this table (1.5) is close to the value that maximizes the likelihood, but the value I chose for the mean (6.5) is probably too small—the maximum looks to occur when the mean is around 7.5.

Figure 4.1 also illustrates two potential problems with numerical methods to calculate the likelihood. If your model has many parameters, drawing a graph becomes hard, and more importantly, the parameter space might become difficult to explore: you might think you have a maximum in one region, but you might miss another peak somewhere else in the high-dimensional space. The second problem is that you have to choose individual points in the parameter space to numerically evaluate the likelihood—there might always be another set of parameters between the points you evaluated that has a slightly higher likelihood. Numerical optimization of objective functions in machine learning is a major area of current research, and we will only scratch the surface in this book.

4.4 HOW TO MAXIMIZE THE LIKELIHOOD ANALYTICALLY

Although numerical approaches are always possible nowadays, it's still faster (and more fun!) to find the exact mathematical maximum of the likelihood function, if you can. We'll derive the MLEs for the univariate Gaussian likelihood introduced earlier. The problem is complicated enough to illustrate the major concepts of likelihood, as well as some important mathematical notations and tricks that are widely used to solve statistical modeling and machine learning problems. Once we have written down the likelihood function, the next step is to find the maximum of this function by taking the derivatives with respect to the parameters, setting them equal to zero, and solving for the maximum likelihood estimators. Needless to say, this probably seems very daunting at this point. But if you make it through this book, you'll look back at this problem with fondness because it was so *simple* to find the analytic solutions. The mathematical trick that makes this problem go from looking very hard to being relatively easy to solve is the following: take the logarithm. Instead of working with likelihoods, in practice, we'll almost always use log-likelihoods because of their mathematical convenience. (Log-likelihoods are also easier to work with numerically because instead of very small positive numbers near zero, we can work with big negative numbers.) Because the logarithm is monotonic (it doesn't change the ranks of numbers), the maximum of the log-likelihood is also the maximum of the likelihood. So here's the mathematical magic:

$$\log L = \log \prod_{i=1}^{n} N(X_i \mid \mu, \sigma) = \sum_{i=1}^{n} \log N(X_i \mid \mu, \sigma)$$

$$= \sum_{i=1}^{n} -\log \sigma - \frac{1}{2}\log(2\pi) - \frac{(X_i - \mu)^2}{2\sigma^2}$$

In the equations above, I have used several properties of the logarithm: $\log(1/x) = -\log x$, $\log(xy) = \log(x) + \log(y)$, and $\log(e^x) = x$. This formula for the log-likelihood might not look much better, but remember that we are trying to find the parameters that maximize this function. To do so, we want to take its derivative with respect to the parameters and set it equal to zero. To find the MLE of the mean, μ, we will take derivatives with respect to μ. Using the linearity of the derivative operator, we have

$$\frac{\partial}{\partial \mu} \log L = \frac{\partial}{\partial \mu} \sum_{i=1}^{n} -\log \sigma - \frac{1}{2}\log(2\pi) - \frac{(X_i - \mu)^2}{2\sigma^2}$$

$$= \sum_{i=1}^{n} -\frac{\partial}{\partial \mu}\log \sigma - \frac{1}{2}\frac{\partial}{\partial \mu}\log(2\pi) - \frac{\partial}{\partial \mu}\frac{(X_i - \mu)^2}{2\sigma^2} = 0$$

Since two of the terms have no dependence on μ, their derivatives are simply zero. Taking the derivatives, we get

$$\frac{\partial}{\partial \mu} \log L = \sum_{i=1}^{n} -0 - 0 + \frac{2(X_i - \mu)}{2\sigma^2} = \frac{1}{\sigma^2} \sum_{i=1}^{n} (X_i - \mu) = 0$$

where in the last step I took out of the sum the σ^2 that didn't depend on i. Since we can multiply both sides of this equation by σ^2, we are left with

$$\sum_{i=1}^{n}(X_i - \mu) = \sum_{i=1}^{n} X_i - \sum_{i=1}^{n} \mu = \sum_{i=1}^{n} X_i - n\mu = 0$$

which we can actually solve by

$$\mu = \mu_{MLE} = \frac{1}{n} \sum_{i=1}^{n} X_i = m_X$$

This equation tells us the value of μ that we should choose if we want to maximize the likelihood. I hope that it is clear that the suggestion is simply to choose the sum of the observations divided by the total number of observations—in other words, the average. I have written μ_{MLE} to remind us that this is the maximum likelihood estimator for the parameter, rather than the parameter itself.

Notice that although the likelihood function (illustrated in Figure 4.1) depends on both parameters, the formula we obtained for the μ_{MLE} doesn't. A similar (slightly more complicated) derivation is also possible for the standard deviation:

$$\frac{\partial}{\partial \sigma} \log L = 0 \rightarrow \sigma_{MLE} = \sqrt{\frac{1}{n} \sum_{i=1}^{n} (X_i - \mu)^2} = s_X$$

In the MLE for the standard deviation, there is an explicit dependence on the mean. Because in order to maximize the likelihood, the derivatives with respect to *all* the parameters must be zero, to get the MLE for the standard deviation, you need to first calculate the MLE for the mean and plug it in to the formula for the MLE of the standard deviation.

In general, setting the derivatives of the likelihood with respect to all the parameters to zero leads to a set of equations with as many equations and unknowns as the number of parameters. In practice, there are few problems of this kind that can be solved analytically.

THE DISTRIBUTION OF PARAMETER ESTIMATES FOR MLEs

Assuming that you have managed to maximize the likelihood of your model using either analytic or numerical approaches, it is sometimes possible to take advantage of the very well-developed statistical theory in this area to do hypothesis testing on the parameters. The maximum likelihood estimator is a function of the data, and therefore it will not give the same answer if another random sample is taken from the distribution. However, it is known that (under certain assumptions) the MLEs will be Gaussian distributed, with means equal to the true means of the parameters, and variances related to the second derivatives of the likelihood at the maximum, which are summarized in the so-called Fisher Information matrix (which I abbreviate as FI).

$$\mathrm{Var}(\theta_{MLE}) = E[-FI^{-1}]_{\theta=\theta_{MLE}}$$

This formula says that the variance of the parameter estimates is the (1) expectation of the negative of (2) inverse of the (3) Fisher information matrix evaluated at the maximum of the likelihood (so that all parameters have been set equal to their MLEs). I've written the numbers to indicate that getting the variance of the parameter estimates is actually a tedious three-step process, and it's rarely used in practice for that reason. However, if you have a simple model, and don't mind a little math, it can be incredibly useful to have these variances. For example, in the case of the Gaussian distribution, there are two parameters (μ and σ), so the Fisher information matrix is a 2×2 matrix.

$$FI = \begin{bmatrix} \dfrac{\partial^2 \log L}{\partial \mu^2} & \dfrac{\partial^2 \log L}{\partial \mu \partial \sigma} \\ \dfrac{\partial^2 \log L}{\partial \mu \partial \sigma} & \dfrac{\partial^2 \log L}{\partial \sigma^2} \end{bmatrix} = \begin{bmatrix} -\dfrac{n}{\sigma^2} & 0 \\ 0 & -\dfrac{2n}{\sigma^2} \end{bmatrix}$$

The first step in getting the variance of your estimator is evaluating these derivatives. In most cases, this must be done numerically, but in textbook examples they can be evaluated analytically. For the Gaussian model, at the maximum of the likelihood they have the simple formulas that I've given here.

The second derivatives measure the change in the slope of the likelihood function, and it makes sense that they come up here because the variance of the maximum likelihood estimator is related intuitively to the shape of the likelihood function near the maximum. If the likelihood surface is very flat around the estimate, there is less certainty, whereas if the MLE is at a very sharp peak in the likelihood surface, there is a lot of certainty—another sample from the same data is likely to give nearly the same maximum. The second derivatives measure the local curvature of the likelihood surface near the maximum.

Once you have the derivatives (using the values of the parameters at the maximum), the next step is to invert this matrix. In practice, this is not possible to do analytically for all but the simplest statistical models. For the Gaussian case, the matrix is diagonal, so the inverse is just

$$FI^{-1} = \begin{bmatrix} -\dfrac{\sigma^2}{n} & 0 \\ 0 & -\dfrac{\sigma^2}{2n} \end{bmatrix}$$

Finally, once you have the inverse, you simply take the negative of the diagonal entry in the matrix that corresponds to the parameter you're interested in, and then take the expectation. So the variance for the mean would be σ^2/n. This means that the distribution of μ_{MLE} is Gaussian, with mean equal to the true mean, and standard deviation is equal to the true standard deviation divided by the square root of n.

You probably noticed that this calculation also tells us the distribution for the MLE of the variance. Since the estimate of the variance can only be positive, it's in some sense surprising that statistical theory says that it should have a Gaussian distribution (which we know gives probabilities to negative numbers). The resolution of this contradiction is that one of the assumptions under which the MLEs approach Gaussian distributions is that the sample size is very large, which limits the applicability of the theory in many practical situations. For small sample sizes, the distribution of the variance estimate is not very Gaussian at all. In fact, it has a nonstandard distribution.

4.5 OTHER OBJECTIVE FUNCTIONS

Despite the popularity, conceptual clarity and theoretical properties of maximum likelihood estimation, there are (many) other objective functions and corresponding estimators that are widely used.

Another simple, intuitive objective function is the "least squares"— simply adding up the squared differences between the model and the data. Minimizing the sum of squared differences leads to the maximum likelihood estimates in many cases, but not always. One good thing about least squares estimation is that it can be applied even when your model doesn't actually conform to a probability distribution (or it's very hard to write out or compute the probability distribution).

One of the most important objective functions for machine learning is the so-called posterior probability and the corresponding Maximum-Apostiori-Probability or MAP estimates/estimators. In contrast to ML estimation, MAP estimation says: "choose the parameters so the model is most probable given the data we observed." Now, the objective function is $P(\theta|X)$ and the equation to solve is

$$\frac{\partial}{\partial\theta} P(\theta|X) = 0$$

As you probably already guessed, the MAP and ML estimation problems are related via Bayes' theorem, so that this can be written as

$$P(\theta|X) = \frac{P(\theta)}{P(X)} P(X|\theta) = \frac{P(\theta)}{P(X)} L$$

Once again, it is convenient to think about the optimization problem in log space, where the objective function breaks into three parts, only two of which actually depend on the parameters.

$$\frac{\partial}{\partial\theta} \log P(\theta|X) = \frac{\partial}{\partial\theta} \log P(\theta) + \frac{\partial}{\partial\theta} \log L - \frac{\partial}{\partial\theta} \log P(X)$$

$$= \frac{\partial}{\partial\theta} \log P(\theta) + \frac{\partial}{\partial\theta} \log L = 0$$

Interestingly, optimizing the posterior probability therefore amounts to optimizing the likelihood function *plus* another term that depends only on the parameters.

The posterior probability objective function turns out to be one of a class of so-called "penalized" likelihood functions where the likelihood is combined with mathematical functions of the parameters to create a new objective function. As we shall see in Chapter 9, these objective functions turn out to underlie several intuitive and powerful machine learning methods that we will see in later chapters.

Some very practically important classes of models used for machine learning (e.g., neural networks and SVMs) have specialized objective functions that have been developed for them. These models do not have probabilistic interpretations, so their objective functions cannot usually be related to likelihoods. Nevertheless, these models still have parameters to estimate (or "train"), and the efficiency and accuracy of the "training" algorithms available is critical to the practical applicability of these methods.

No matter what objective function is chosen, estimation usually always involves solving a mathematical optimization problem, and in practice this is almost always done using a computer—either with a statistical software package such as R or MATLAB®, or using purpose-written codes.

BIAS, CONSISTENCY, AND EFFICIENCY OF ESTIMATORS

In order to facilitate debates about which estimators are the best, statisticians developed several objective criteria that can be used to compare estimators. For example, it is commonly taught that the ML estimator of the standard deviation for the Gaussian is biased. This means that for a sample of data, the value of the standard deviation obtained using the formula given will tend to (in this case) underestimate the "true" standard deviation of the values in the pool. On the other hand, the estimator is consistent, meaning that, as the sample size drawn from the pool approaches infinity, the estimator does converge to the "true" value. Finally, the efficiency of the estimator describes how quickly the estimate approaches the truth as a function of the sample size. In modern molecular biology, these issues will almost always be taken care of by the computer statistics package used to do the calculations. Thus, although this is traditionally covered at length in introductory statistics courses, it's something we rarely have to worry about in computer-assisted science.

So how do we choose an objective function? In practice, as biologists we usually choose the one that's simplest to apply, where we can find a way to reliably optimize it. We're not usually interested in debating about whether the likelihood of the model is more important than the likelihood of the data. Instead, we want to know something about the parameters that are being estimated—testing our hypothesis about whether this cell line yields greater protein abundance than another cell line, whether a sample is a breast tumor and not healthy tissue, whether there are two groups of interbreeding populations or three, or how well mRNA levels predict protein abundance. So as long as we use the same method of estimation on all of our data, it's probably not that important which estimation method we use.

BAYESIAN ESTIMATION AND PRIOR DISTRIBUTIONS

As we have seen, the MAP objective function and more generally penalized likelihood methods can be related to the ML objective function through the use of Bayes' theorem. For this reason, these methods sometimes are given names with the word "Bayesian" in them. However, as long as a method results in a single estimator for parameters, it is not really Bayesian in spirit. Truly Bayesian estimation means that you don't try to pin down a single value for your parameters. Instead, you embrace the fundamental uncertainty that any particular estimate for your parameters is just one possible estimate drawn from a pool. True Bayesian statistics mean that you consider the entire distribution of your parameters, given your data and your prior beliefs about what the parameters should be. In practice, Bayesian estimation is not usually used in biology, because biologists want to know the values of their parameters. We don't usually want to consider the whole distribution of expression levels of our gene that are compatible with the observed data: we want to know the level of the gene.

One interesting example where the Bayesian perspective of estimating probability distributions matches our biological intuition and has led to a widely used data analysis approach is in the problem of population structure inference. Population structure is the situation where the individuals from a species are separated into genetic groups such that recombination is much more common within the genetically separated groups than between them. Population structure is often caused by geographical constraints: western, central, and eastern chimpanzees are separated from each other by rivers. The individuals in each population rarely cross, and therefore there is little "gene flow" between populations. Understanding population structure is important for both evolutionary genetics and human medical genetics. For example, when a case-control study is performed to identify loci associated with a disease, if the population structure of the cases is different than that of the controls, spurious associations will be identified. In general, for a sample of humans, population history is not always known, but population structure can be inferred based on differences in allele frequencies between populations, and individuals can be assigned to populations based on their genotypes. In the simplest case, where there is no recombination at all between subpopulations, individuals should be assigned to exactly one subpopulation. However, when rare recombination events do occur, so-called "admixed" individuals may appear in a population sample. Admixture simply means that individuals have parents (or other more distant ancestors) from more than one of the subpopulations. Since the assortment of genetic material into offspring is random, these admixed individuals will truly be drawn from more than one subpopulation. In this case, the Bayesian perspective says, rather than trying to assign each individual to a population, estimate the probability distribution over the ancestral subpopulations. This Bayesian approach is implemented in the widely used software STRUCTURE

(Pritchard et al. 2000) and allows an individual to be partially assigned to several subpopulations.

Although Bayesian estimation is not used that often for molecular biology data analysis, the closely associated concepts of prior and posterior distributions are very powerful and widely used. Because the Bayesian perspective is to think of the parameters as random variables that need to be estimated, models are used to describe the distributions of the parameters both before (prior) and after considering the observations (posterior). Although it might not seem intuitive to think that the parameters have a distribution before we consider the data, in fact it makes a lot of sense: we might require the parameters to be between zero and infinity if we are using a Poisson model for numbers of sequence reads or ensure that they add up to one if we are using a multinomial or binomial model for allele frequencies in a population. The idea of prior distributions is that we can generalize this to quantitatively weigh the values of the parameters by how likely they might turn out to be. Of course, if we don't have any prior beliefs about the parameters, we can always use uniform distributions, so that all possible values of the parameters are equally likely (in Bayesian jargon, these are called uninformative priors). However, as we shall see, we will find it too convenient to resist putting biological knowledge into our models using priors.

4.6 MULTIVARIATE STATISTICS

An important generalization of the statistical models that we've seen so far is to the case where multiple events are observed at the same time. In the models we've seen so far, observations were single events: yes or no, numbers or letters. In practice, a modern molecular biology experiment typically measures more than one thing, and a genomics experiment might yield measurements for thousands of things: for example, measurements for all the genes in the genome.

A familiar example of an experiment of this kind might be a set of genome-wide expression level measurements. In the ImmGen data, for each gene, we have measurements of gene expression over ~200 different cell types. Although in the previous chapters we considered each of the cell types independently, a more comprehensive way to describe the data is that for each gene, the observation is actually a vector, X, of length ~200, where each element of the vector is the expression measurement for a specific cell type. Alternatively, it might be more convenient for other questions to think of each observation as a cell type, where the observation is now a vector of 24,000 gene expression measurements. This situation is known in statistics as "multivariate" to describe the idea that multiple variables are being measured simultaneously. Conveniently, the familiar Gaussian distribution generalizes to the multivariate case, except the single numbers (scalar) mean and variance parameters are now replaced with a mean vector and (co)variance matrix:

$$N\left(\vec{X} \mid \vec{\mu}, \boldsymbol{\Sigma}\right) = \frac{1}{\sqrt{|\boldsymbol{\Sigma}|(2\pi)^d}} e^{-(1/2)(\vec{X}-\vec{\mu})^T \boldsymbol{\Sigma}^{-1}(\vec{X}-\vec{\mu})}$$

In the formula for the multivariate Gaussian, I've used a small d to indicate the dimensionality of the data, so that d is the length of the vectors μ and X, and the covariance is a matrix of size $d \times d$. In this formula, I've explicitly written small arrows above the vectors and bolded the matrices. In general, the machine learning people will not do this (I will adopt their convention) and it will be left to the reader to keep track of what are the scalars, vectors, and matrices. If you're hazy on your vector and matrix multiplications and transposes, you'll have to review them in order to follow the rest of this section (and most of this book).

A QUICK REVIEW OF VECTORS, MATRICES, AND LINEAR ALGEBRA

As I've already mentioned, a convenient way to think about multiple observations at the same time is to think about them as lists of numbers, which are known as vectors in mathematical jargon. We will refer to the list of numbers X as a vector $X = (x_1, x_2, x_3, \ldots, x_n)$, where n is the "length" or dimensionality of the vector (the number of numbers in the list). Once we've defined these lists of numbers, we can go ahead and define arithmetic and algebra on these lists. One interesting wrinkle to the mathematics of lists is that for any operation we define, we have to keep track of whether the result is actually a list or a number. For example, a simple subtraction of one vector from another looks like

$$X - Y = (x_1, x_2, \ldots, x_n) - (y_1, y_2, \ldots, y_n) = (x_1 - y_1, x_2 - y_2, \ldots, x_n - y_n)$$

which turns out to be a vector. On the other hand, the dot product

$$X \cdot Y = x_1 y_1 + x_2 y_2 + \cdots + x_n y_n = \sum_{i=1}^{n} x_i y_i$$

gives us is a number. The generalization of algebra means that we can write equations like

$$X - Y = 0,$$

which means that

$$(x_1 - y_1, x_2 - y_2, \ldots, x_n - y_n) = (0, 0, \ldots, 0)$$

and is therefore a shorthand way of writing n equations in one line.

Since mathematicians love generalizations, there's no reason we can't generalize the idea of lists to also include a list of lists, so that each element of the list is actually a vector. This type of object is what we call a matrix. $A = (X_1, X_2, X_3, \ldots, X_m)$, where $X_1 = (x_1, x_2, x_3, \ldots, x_n)$. To refer to each element of A, we can write $A_{11}, A_{12}, A_{13}, \ldots, A_{21}, A_{22}, A_{23}, \ldots, A_{mn}$. We can then go ahead and define some mathematical operations on matrices as well: If A and B are matrices, $A - B = C$ means that for all i and j, $C_{ij} = A_{ij} - B_{ij}$.

We can also do mixtures of matrices and vectors and numbers:

$$cx + Ay = \left(cx_1 + \sum_{j=1}^{m} A_{1j} y_j, cx_2 + \sum_{j=1}^{m} A_{2j} y_j, \ldots, cx_n + \sum_{j=1}^{m} A_{nj} y_j \right)$$

where

 c is a number

 x and y are vectors

 A is a matrix

This turns out to be a vector.

However, there's one very inelegant issue with the generalization to matrices: what we mean when we refer to the value A_{ij} depends on whether the i refers to the index in 1 through m or 1 through n. In other words, we have to keep track of the structure of the matrix. To deal with this issue, linear algebra has developed a set of internally consistent notations, which are referred to as the "row" or "column" conventions. So anytime I write the vector, x, by default I mean the "column" vector

$$X = \begin{pmatrix} x_1 \\ x_2 \\ \vdots \\ x_n \end{pmatrix}, \quad X^T = (x_1 \quad x_2 \quad \cdots \quad x_n)$$

To indicate the "row" vector, I have to write the "transpose" of X, or X^T. The transpose is defined as the operation of switching all the rows and columns. So in fact, there are two kinds of products that can be defined:

$$X \cdot Y = X^T Y = \begin{pmatrix} x_1 \\ x_2 \\ \vdots \\ x_n \end{pmatrix} (y_1 \quad y_2 \quad \cdots \quad y_n) = \sum_{i=1}^{n} x_i y_i,$$

which is the familiar dot product also known as the "inner product," and produces a number, while

$$XY^T = (x_1 \quad x_2 \quad \cdots \quad x_n)\begin{pmatrix} y_1 \\ y_2 \\ \vdots \\ y_m \end{pmatrix} = \begin{pmatrix} x_1y_1 & x_2y_1 & \cdots & x_ny_1 \\ x_1y_2 & x_2y_2 & & \\ \vdots & & \ddots & \\ x_1y_m & & & x_ny_m \end{pmatrix}$$

which is the so-called "outer" product that takes two vectors and produces a matrix.

Although you don't really have to worry about this stuff unless you are doing the calculations, I will try to use consistent notation, and you'll have to get used to seeing these linear algebra notations as we go along.

Finally, an interesting point here is to consider the generalization beyond lists of lists: it's quite reasonable to define a matrix, where each element of the matrix is actually a vector. This object is called a tensor. Unfortunately, as you can imagine, when we get to objects with three indices, there's no simple convention like "rows" and "columns" that we can use to keep track of the structure of the objects. I will at various times in this book introduce objects with more than two indices—especially when dealing with sequence data. However, in those cases, it won't be obvious what the generalizations of addition, subtraction from linear algebra mean exactly, because we won't be able to keep track of the indices. Sometimes, I will try to use these and things might get confusing, but in many cases I'll have to write out the sums explicitly when we get beyond two indices.

Matrices also have different types of multiplications: the matrix product produces a matrix, but there are also inner products and outer products that produce other objects. A related concept that we've already used is the "inverse" of a matrix. The inverse is the matrix that multiplies to give a matrix with 1's along the diagonal (the so-called identity matrix, I).

$$AA^{-1} = I = \begin{pmatrix} 1 & 0 & \cdots & 0 \\ 0 & 1 & & \\ \vdots & & \ddots & \\ 0 & & & 1 \end{pmatrix}$$

Although this might all sound complicated, multivariate statistics is easy to understand because there's a very straightforward, beautiful geometric interpretation to it. The idea is that we think of each component of the observation vector (say, each gene's expression level in a specific cell type) as a "dimension." If we measure the expression level of two genes in each cell type, we have two-dimensional data. If we measure three genes, then we have three-dimensional data. If 24,000 genes, then we have ... you get the idea. Of course, we won't have an easy time making graphs of 24,000-dimensional space, so we'll typically use two- or three-dimensional examples for illustrative purposes. Figure 4.2 tries to illustrate the idea.

In biology, there are lots of other types of multivariate data. For example, one might have observations of genotypes and phenotypes for a sample of individuals. Another ubiquitous example is DNA sequences: the letter

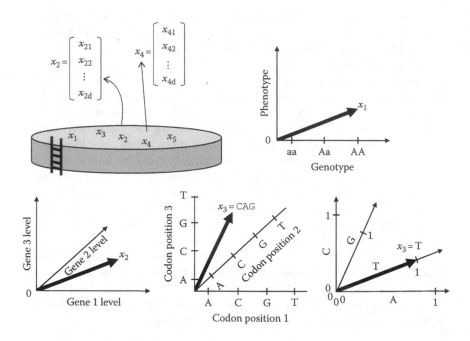

FIGURE 4.2 Multivariate observations as vectors. Different types of data typically encountered in molecular biology can be represented as lists or vectors. The top left shows i.i.d. multivariate observations in a pool. Each observation corresponds to a vector (list of observations) of length d. On the bottom right, a single gene expression observation for three genes is represented as a vector in three-dimensional space. On the top right, a two-dimensional observation of phenotype and genotype is indicated in a space with a discrete horizontal dimension and a continuous vertical dimension. On the bottom center, a codon is represented as a point in three-dimensional space where each dimension corresponds to one codon position. In the bottom right, a single nucleotide position is represented in a four-dimensional space. Note that discrete observations like sequences or genotypes can be represented mathematically in different ways.

at each position can be thought of as one of the dimensions. In this view, each of our genomes represents three billion dimensional vectors sampled from the pool of the human population. In an even more useful representation, each position in a DNA (or protein) sequence can be represented as a 4 (or 20)-dimensional vector, and the human genome can be thought of as a 3 billion × 4 matrix of 1s and 0s. In these cases, the components of observations are not all numbers, but this should not stop us from using the geometrical interpretation that each observation is a vector in a high-dimensional space (Figure 4.2 illustrates multivariate data).

A key generalization that becomes available in multivariate statistics is the idea of correlation. Although we will still assume that the observations are i.i.d., the dimensions are not necessarily independent (Figure 4.3). For example, in a multivariate Gaussian model for cell-type gene expression, the observation of a highly expressed gene X might make us more likely to observe a highly expressed gene Y. In the multivariate Gaussian model, the correlation between the dimensions is controlled by the off-diagonal elements in the covariance matrix, where each off-diagonal entry summarizes the correlation between a pair of dimensions (Figure 4.3). Intuitively, an off-diagonal term of zero implies that there is no correlation between two dimensions. In a multivariate Gaussian model where all the dimensions are independent, the off-diagonal terms of the covariance matrix are all zero, so the covariance is said to be diagonal. A diagonal covariance leads to a symmetric, isotropic, or (most confusingly) "spherical" distribution.

4.7 MLEs FOR MULTIVARIATE DISTRIBUTIONS

The ideas that we've already introduced about optimizing objective functions can be transferred directly from the univariate case to the multivariate case. Only minor technical complication will arise because multivariate distributions have more parameters, and therefore the set of equations to solve can be larger and more complicated.

To illustrate the kind of mathematical tricks that we'll need to use, we'll consider two examples of multivariate distributions that are very commonly used. First, the multinomial distribution, which is the multivariate generalization of the binomial. This distribution describes the numbers of times we observe events from multiple categories. For example, the traditional use of this type of distribution would be to describe the number of times each face of a die (1–6) turned up. In bioinformatics, it is often used to describe the numbers of each of the bases in DNA (A, C, G, T).

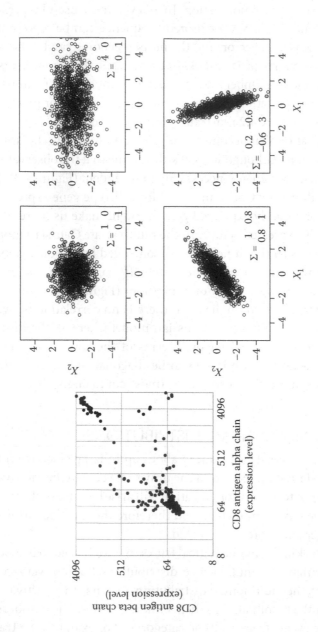

FIGURE 4.3   Multivariate Gaussians and correlation. The panel on the left shows real gene-expression data for the CD8 antigen (from ImmGen). The panel on the right shows four parameterizations of the multivariate Gaussian in two dimensions. In each case, the mean is at (0, 0). Notice that none of the simple Gaussian models fit the observed CD8 expression data very well.

If we say the X is the number of counts of the four bases, and f is a vector of probabilities of observing each base, such that $\sum_i f_i = 1$, where i indexes the four bases. The multinomial probability for the counts of each base in DNA is given by

$$MN(X|f) = \frac{(X_A + X_C + X_G + X_T)!}{X_A!X_C!X_G!X_T!} f_A^{X_A} f_C^{X_C} f_G^{X_G} f_T^{X_T}$$

$$= \frac{\left(\sum_{i \in \{A,C,G,T\}} X_i\right)!}{\prod_{i \in \{A,C,G,T\}} X_i!} \prod_{i \in \{A,C,G,T\}} f_i^{X_i}$$

The term on the right-hand side (with the factorial of the sum over the product of factorials) has to do with the number of "ways" that you could have observed, say, $X = (535\ 462\ 433\ 506)$ for A, C, G, and T. For our purposes (to derive the MLEs), we don't need to worry about this term because it doesn't depend on the parameters, so when you take the log and then the derivative, it will disappear. We start by writing the log-likelihood of the data given the model

$$\log L = \log MN(X|f) = \log \frac{\left(\sum_{i \in \{A,C,G,T\}} X_i\right)!}{\prod_{i \in \{A,C,G,T\}} X_i!} + \log \prod_{i \in \{A,C,G,T\}} f_i^{X_i}$$

The parameters are the f's, so the equation to solve for each one is

$$\frac{\partial \log L}{\partial f_A} = \frac{\partial}{\partial f_A} \sum_{i \in \{A,C,G,T\}} X_i \log f_i = 0$$

If we solved this equation directly, we get

$$\frac{\partial \log L}{\partial f_A} = \frac{X_A}{f_A} = 0$$

Although this equation seems easy to solve, there is one tricky issue: the sum of the parameters has to be 1. If we solved the equation here, we'd always set each of the parameters to infinity, in which case they would not sum up to one. The optimization (taking derivatives and setting them to zero) doesn't know that we're working on a probabilistic model—it's just trying to do the optimization. In order to enforce that the parameters stay between 0 and 1, we need to add a constraint to the optimization. This is most easily done through the method of Lagrange multipliers, where we rewrite the constraint as an equation that equals zero, for example, $1 - \sum_i f_i = 0$, and add it to the function we are trying to optimize multiplied by a constant, the so-called Lagrange multiplier, λ:

$$\frac{\partial \log L}{\partial f_A} = \frac{\partial}{\partial f_A} \sum_{i \in \{A,C,G,T\}} X_i \log f_i + \lambda \left(1 - \sum_{i \in \{A,C,G,T\}} f_i \right) = 0$$

Taking the derivatives gives

$$\frac{\partial \log L}{\partial f_A} = \frac{X_A}{f_A} - \lambda = 0$$

which we can solve to give

$$(f_A)_{MLE} = \frac{X_A}{\lambda}$$

Here, I have used MLE to indicate that we now have the maximum likelihood estimator for the parameter f_A. Of course, this is not very useful because it is in terms of the Lagrange multiplier. To figure out what the actual MLEs are, we have to think about the constraint $1 - \sum_i f_i = 0$. Since we need the derivatives with respect to all the parameters to be 0, we'll get a similar equation for f_C, f_G, and f_T. Putting these together gives us

$$\sum_{i \in \{A,C,G,T\}} (f_i)_{MLE} = \sum_{i \in \{A,C,G,T\}} \frac{X_i}{\lambda} = 1$$

or

$$\lambda = \sum_{i \in \{A,C,G,T\}} X_i$$

which says that λ is just the total number of bases we observed. So the MLE for the parameter f_A is just

$$f_A = \frac{X_A}{\sum_{i \in \{A,C,G,T\}} X_i}$$

where I have now dropped the cumbersome MLE notation. This formula is the intuitive result that the estimate for probability of observing A is just the fraction of bases that were actually A.

THE MULTINOMIAL DISTRIBUTION AND THE CATEGORICAL DISTRIBUTION

As I said earlier, the multinomial distribution is the multivariate analogue of the binomial distribution (the classical distribution for describing the number of heads or tails after n tries). We can also obtain the multinomial distribution in a similar way, by considering the number of outcomes from, say, a die with m sides. Each throw of the die is described by the "discrete" or "categorical" distribution:

$$P(X|f) = \prod_{j=1}^{m} f_i^{X_j} \quad f_j > 0, \quad \sum_{j=1}^{m} f_j = 1, \quad X_j \in \{0,1\}, \quad \sum_{j=1}^{m} X_j = 1$$

This is the multivariate generalization of the Bernoulli distribution. It gives the probability for a single roll of a die with m sides. The notation for this gets a bit tedious, but hope it's clear that I'm now using a vector, f, to contain the probabilities of each type of event occurring, and a vector X to represent the m possible outcomes. In this distribution, the components of X must sum to 1 because only one outcome (of m) is allowed each time we throw the die.

After n observations, we get the multinomial distribution:

$$P(X|f,n) = \frac{n!}{\prod_{j=1}^{m} \left(\sum_{i=1}^{n} X_{ij} \right)!} \prod_{j=1}^{m} f_j^{\sum_{i=1}^{n} X_{ij}}$$

Notice that this formula looks slightly more complicated notation than the one I used when deriving the MLEs, but now I've kept the notation consistent with what I used for the categorical distribution. Similar to the previous equation

$$f_j > 0, \quad \sum_{j=1}^{m} f_j = 1, \quad X_j \in \{0,1\}, \quad \sum_{j=1}^{m} X_j = 1$$

I hope it's clear that in the multinomial distribution formula, $\sum_{i=1}^{n} X_{ij}$ is just the number of observations of each type. The notation I've used here, where each observation can either be 0 or 1 for each type (instead of the numbers of each type of observation) sometimes makes derivations of MLEs and various algorithms a bit easier.

A more complicated example is to find the MLEs for the multivariate Gaussian. We'll start by trying to find the MLEs for the mean. As before, we can write the log-likelihood

$$\log L = \log P(X \mid \theta) = \log \prod_{i=1}^{n} N(X_i \mid \mu, \Sigma)$$

$$= \sum_{i=1}^{n} -\frac{1}{2} \log\left[|\Sigma| (2\pi)^d \right] - \frac{1}{2}(X_i - \mu)^T \Sigma^{-1}(X_i - \mu)$$

Since we are working in the multivariate case, we now need to take a derivative with respect to a vector μ. One way to do this is to simply take the derivative with respect to each component of the vector. So for the first component of the vector, we could write

$$\frac{\partial \log L}{\partial \mu_1} = \sum_{i=1}^{n} \frac{\partial}{\partial \mu_1}\left[-\frac{1}{2}\sum_{j=1}^{d}(X_{ij} - \mu_j)\sum_{k=1}^{d}(\Sigma^{-1})_{jk}(X_{ik} - \mu_k) \right] = 0$$

$$\frac{\partial \log L}{\partial \mu_1} = \sum_{i=1}^{n} \frac{\partial}{\partial \mu_1}\left[-\frac{1}{2}\sum_{j=1}^{d}\sum_{k=1}^{d}(X_{ij} - \mu_j)(\Sigma^{-1})_{jk}(X_{ik} - \mu_k) \right] = 0$$

$$\frac{\partial \log L}{\partial \mu_1} = \sum_{i=1}^{n} \frac{\partial}{\partial \mu_1}\left[-\frac{1}{2}\sum_{j=1}^{d}\sum_{k=1}^{d}(\Sigma^{-1})_{jk}(X_{ij}X_{ik} - \mu_j X_{ik} - \mu_k X_{ij} + \mu_j \mu_k) \right] = 0$$

where I have tried to write out the matrix and vector multiplications explicitly. Since the derivative will be zero for all terms that don't depend on the first component of the mean, we have

$$\frac{\partial \log L}{\partial \mu_1} = -\frac{1}{2} \sum_{i=1}^{n} \frac{\partial}{\partial \mu_1} \Bigg[(\Sigma^{-1})_{11}(-\mu_1 X_{i1} - \mu_1 X_{i1} + \mu_1 \mu_1)$$

$$+ \sum_{j=2}^{d} (\Sigma^{-1})_{j1}(-\mu_j X_{i1} - \mu_1 X_{ij} + \mu_j \mu_1)$$

$$+ \sum_{k=2}^{d} (\Sigma^{-1})_{1k}(-\mu_1 X_{i1} - \mu_k X_{i1} + \mu_1 \mu_k) \Bigg] = 0$$

Because of the symmetry of the covariance matrix, the last two terms are actually the same:

$$\frac{\partial \log L}{\partial \mu_1} = -\frac{1}{2} \sum_{i=1}^{n} \frac{\partial}{\partial \mu_1} \Bigg[(\Sigma^{-1})_{11}(-\mu_1 X_{i1} - \mu_1 X_{i1} + \mu_1 \mu_1)$$

$$+ 2 \sum_{j=2}^{d} (\Sigma^{-1})_{j1}(-\mu_j X_{i1} - \mu_1 X_{ij} + \mu_j \mu_1) \Bigg] = 0$$

Differentiating the terms that do depend on the first component of the mean gives

$$\frac{\partial \log L}{\partial \mu_1} = -\frac{1}{2} \sum_{i=1}^{n} \Bigg[(\Sigma^{-1})_{11}(-2X_{i1} + 2\mu_1) + 2 \sum_{j=2}^{d} (\Sigma^{-1})_{j1}(-X_{ij} + \mu_j) \Bigg] = 0$$

$$\frac{\partial \log L}{\partial \mu_1} = -\sum_{i=1}^{n} \Bigg[(\Sigma^{-1})_{11}(-X_{i1} + \mu_1) + \sum_{j=2}^{d} (\Sigma^{-1})_{j1}(-X_{ij} + \mu_j) \Bigg] = 0$$

Merging the first term back in to the sum, we have

$$\frac{\partial \log L}{\partial \mu_1} = -\sum_{i=1}^{n} \Bigg[\sum_{j=1}^{d} (-X_{ij} + \mu_j)(\Sigma^{-1})_{j1} \Bigg] = -\sum_{i=1}^{n} (\mu - X_i)^T (\Sigma^{-1})_1 = 0$$

where I have abused the notation somewhat to write the first column of the inverse covariance matrix as a vector, $(\Sigma^{-1})_1$. Notice the problem: although were trying to find the MLE for the first component of the mean only, we have an equation that includes all the components of the mean through the off-diagonal elements of the covariance matrix. This means we have a single equation with d variables and unknowns, which obviously cannot be solved uniquely. However, when we try to find the maximum likelihood parameters, we have to set *all* the derivatives with respect to all the parameters to zero, and we will end up at equations like this for each component of the mean. This implies that we will actually have a set of d equations for each of the d components of the mean, each involving a different row of the covariance matrix. We'll get a set of equations like

$$\frac{\partial \log L}{\partial \mu_2} = \sum_{i=1}^{n} (\mu - X_i)^T (\Sigma^{-1})_2 = 0$$

$$\cdots$$

$$\frac{\partial \log L}{\partial \mu_d} = \sum_{i=1}^{n} (\mu - X_i)^T (\Sigma^{-1})_d = 0$$

(where I multiplied both sides by –1 so as to not bother with the negative sign). We can write the set of equations as

$$\frac{\partial \log L}{\partial \mu} = \sum_{i=1}^{n} (\mu - X_i)^T \Sigma^{-1} = 0$$

where the 0 is now the vector of zeros for all the components of the mean. To solve this equation, we note that the covariance matrix does not depend on i, so we can simply multiply each term of the sum and the 0 vector by the covariance matrix. We get

$$\frac{\partial \log L}{\partial \mu} = \sum_{i=1}^{n} (\mu - X_i)^T = 0$$

The equation can be solved to give

$$\sum_{i=1}^{n} X_i^T = n\mu^T$$

or

$$\mu^T = \frac{1}{n}\sum_{i=1}^{n} X_i^T$$

which says that the MLEs for the components of the mean are simply the averages of the observations in each dimension.

It turns out that there is a much faster way of solving these types of problems using so-called vector (or matrix) calculus. Instead of working on derivatives of each component of the mean individually, we will use clever linear algebra notation to write all of the equations in one line by using the following identity:

$$\frac{\partial}{\partial x}[x^T Ax] = x^T (A + A^T)$$

where A is any matrix, and the derivative is now a derivative with respect to a whole vector, x. Using this trick, we can proceed directly (remember that for a symmetric matrix like the covariance $A = A^T$)

$$\frac{\partial \log L}{\partial \mu} = \frac{\partial}{\partial \mu}\sum_{i=1}^{n} -\frac{1}{2}(X_i - \mu)^T \Sigma^{-1}(X_i - \mu)$$

$$= \sum_{i=1}^{n}\frac{1}{2}(\mu - X_i)^T 2\Sigma^{-1} = \sum_{i=1}^{n}(\mu - X_i)^T \Sigma^{-1} = 0$$

Similar matrix calculus tricks can be used to find the MLEs for (all the components of) the covariance matrix. If you know that $(\partial/\partial A)\log|A| = (A^{-1})^T$ and $(\partial/\partial A)[x^T Ax] = xx^T$, where again A is a matrix and x is a vector, it's not too hard to find the MLEs for the covariance matrix. In general, this matrix calculus is not something that biologists (or even expert bioinformaticians) will be familiar with, so if you ever have to differentiate your likelihood with respect to vectors or matrices, you'll probably have to look up the necessary identities.

4.8 HYPOTHESIS TESTING REVISITED: THE PROBLEMS WITH HIGH DIMENSIONS

Since we've agreed earlier in this chapter that what biologists are usually doing is testing hypotheses, we usually think much more about our hypothesis tests than about our objective functions. Indeed, as we've seen already, it's even possible to do hypothesis testing without specifying parameters or objective functions (nonparametric tests).

Although I said that statistics has a straightforward generalization to high dimensions, in practice one of the most powerful and useful ideas from hypothesis testing, namely, the P-value, does not generalize very well. This has to do with the key idea that the P-value is the probability of observing something as extreme *or more*. In high-dimensional space, it's not clear which direction the "or more" is in. For example, if you observed three genes' average expression levels (7.32, 4.67, 19.3) and you wanted to know whether this was the same as these genes' average expression levels in another set of experiments (8.21, 5.49, 5.37), you could try to form a three-dimensional test statistic, but it's not clear how to sum up the values of the test statistic that are more extreme than the ones you observed—you have to decide which direction(s) to do the sum. Even if you decide which direction you want to sum up each dimension, performing these multidimensional sums is practically difficult as the number of dimensions becomes large.

The simplest way to deal with hypothesis testing in multivariate statistics is just to do a univariate test on each dimension and pretend they are independent. If any dimension is significant, then (after correcting for the number of tests) the multivariate test must also be significant (Mardia et al. 1976). In fact, that's what we were doing in Chapter 3 when we used Bonferroni to correct the number of tests in the gene set enrichment analysis. Even if the tests are not independent, this treatment is conservative, and in practice, we often want to know in which dimension the data differed. In the case of gene set enrichment analysis, we don't really care whether "something" is enriched—we want to know what exactly the enriched category is.

However, there are some cases where we might not want to simply treat all the dimensions independently. A good example of this might be a time course of measurements or measurements that are related in some natural way, like length and width of an iris petal. If you want to test whether one sample of iris petals is bigger than another, you probably don't want to test whether the length is bigger and then whether the height is bigger. You want to combine both into one test. Another example might be if you've made pairs of observations and you want to test if their ratios are different; but the data include a lot of zeros, so you can't actually form the ratios. One possibility is to create a new test statistic and generate some type of empirical null distribution (as described in the first chapter).

However, another powerful approach is to formulate a truly multivariate hypothesis test: a likelihood ratio test or LRT for short.

Formally,

- The observations (or data) are X_1, X_2, ..., X_n, which we will write as a vector X.

- $H0$ is the null hypothesis, and $H1$ is another hypothesis. The two hypotheses make specific claims about the parameters in each model. For example, $H0$ might state that $\theta = \varphi$, some particular values of the parameters, while $H1$ might state that $\theta \neq \varphi$ (i.e., that the parameters are anything but φ).

- The likelihood ratio test statistic is $-2 \log((p(X|H_0))/(p(X|H_1)))$, where any parameters that are not specified by the hypotheses (so-called free parameters) have been set to their maximum likelihood values. (This means that in order to perform a likelihood ratio test, it is necessary to be able to obtain maximum likelihood estimates, either numerically or analytically.)

- Under the null hypothesis, as the size of the dataset goes to infinity, the distribution of the likelihood ratio test statistic approaches a chi-squared distribution, with degrees of freedom equal to the difference in the number of free parameters between the two hypotheses (Mardia et al. 1976). In the case of the likelihood ratio test, at small sample sizes, the distribution might not be well approximated by chi-squared distribution, so for many common applications of the LRT there are corrections used.

The idea of the likelihood ratio test is that when two hypotheses (or models) describe the same data using different numbers of parameters, the one with more free parameters will always achieve a slightly higher likelihood because it can fit the data better. However, the amazing result is that (if the sample size is large enough) the improvement in fit that is simply due to chance is predicted by the chi-squared distribution (which is always positive). If the model with more free parameters fits the data better than the improvement expected by chance, then we should accept that model.

The likelihood ratio test is an example of class of techniques that are widely used in machine learning to decide if adding more parameters to make a more complex model is "worth it" or if it is "over fitting" the data with more parameters than are really needed. We will see other examples of techniques in this spirit later in this book.

4.9 EXAMPLE OF LRT FOR THE MULTINOMIAL: GC CONTENT IN GENOMES

The LRT is very general and powerful, but this also makes it a bit abstract. To illustrate the use of an LRT to test a specific multivariate hypothesis, let's test for differences in GC content between human chromosomes. I counted the numbers of A, C, G, and T on the human X and Y chromosomes, and obtained the following data (Table 4.2):

Since these are counts of discrete categories (there's no numerical order to A, C, G, and T), if we assume A, C, G, and Ts in the genome are i.i.d., we can model this data using the multinomial distribution. Let's go ahead and estimate the proportions by maximum likelihood using the formulas given earlier (except that I've switched the variable name from X to Z to avoid confusions between the random variable X and the chromosome X)

$$(f_{XA})_{MLE} = \frac{Z_{XA}}{\sum_{i \in \{A,C,G,T\}} Z_{Xi}} = \frac{45,648,952}{151,100,560} = 0.302$$

Similar calculations give the MLEs for the parameters of two chromosomes

$$f_X = (0.302 \quad 0.197 \quad 0.198 \quad 0.303)$$

$$f_Y = (0.299 \quad 0.199 \quad 0.201 \quad 0.301)$$

Obviously, these proportions are similar, but not exactly the same. Are these differences statistically significant? The LRT will help us decide.

The null hypothesis is that the X and Y chromosomes have the same proportions of the four nucleotides, while the alternative hypothesis is that they actually have different proportions. To calculate, the LRT have to maximize the likelihood under the two hypotheses. This means we have to calculate the likelihood (or log-likelihood) when the parameters

TABLE 4.2　Numbers of Each Nucleotide on the Human X and Y Chromosomes

	A	C	G	T
X chromosome	45,648,952	29,813,353	29,865,831	45,772,424
Y chromosome	7,667,625	5,099,171	5,153,288	7,733,482

are the MLEs. In this case, it's easiest to start with the alternative hypothesis: we believe that the X and Y chromosomes have different proportions of the four bases, so we allow each chromosome to have four different parameters, so the MLEs are just the proportions we calculated earlier.

Under the null hypothesis, we believe that the two chromosomes have the same proportions of nucleotides, so that a single set of parameters can explain all of the data. We estimate these simply by pooling all of the data:

$$(f_{0A})_{MLE} = \frac{Z_{XA} + Z_{YA}}{\sum_{i \in \{A,C,G,T\}} Z_{Xi} + Z_{Yi}} = \frac{45{,}648{,}952 + 7{,}667{,}625}{151{,}100{,}560 + 25{,}653{,}566} = 0.302$$

$$f_0 = (0.302 \quad 0.198 \quad 0.198 \quad 0.303)$$

where I've used the 0 to indicate that these are the parameters under the null hypothesis $H0$. We can now go ahead and calculate the log-likelihood ratio, which is (see Exercises)

$$\log LR = \log \frac{p(data|H0)}{p(data|H1)} = \log \frac{p(data|f_0)}{p(data|f_X, f_Y)}$$

$$= \log \frac{\prod_{i \in \{A,C,G,T\}} f_{0i}^{Z_{Xi} + Z_{Yi}}}{\prod_{i \in \{A,C,G,T\}} f_{Xi}^{Z_{Xi}} \prod_{i \in \{A,C,G,T\}} f_{Yi}^{Z_{Yi}}}$$

Do your best to navigate the notation here. I'm using Z to represent the numbers of nucleotides actually observed. In practice, we never want to try to calculate $0.3^{45{,}648{,}952}$, we'll actually carry the log through and convert the products to sums and the exponents to products:

$$\log LR = \sum_{i \in \{A,C,G,T\}} (Z_{Xi} + Z_{Yi}) \log f_{0i} - \sum_{i \in \{A,C,G,T\}} Z_{Xi} \log f_{Xi}$$

$$- \sum_{i \in \{A,C,G,T\}} Z_{Yi} \log f_{Yi} \approx -1147.6$$

We can now go ahead and compute the LRT test statistic, which is just -2 times this, or about 2295. We now have to look up this value of the test statistic in the null distribution, which is the chi square, with df equal to the difference in the number of free parameters between the two hypotheses.

It can be a bit tricky to figure this out: for example, it looks like we estimated four parameters for our null hypothesis and eight parameters for our alternative hypothesis (the f's in each case). However, as I mentioned earlier, the parameters of a multinomial have to add up to 1, so in fact, once we choose three of them, the fourth one is automatically specified. Therefore, in this case, there are really only three free parameters in our null hypothesis and six free parameters in the alternative hypothesis. As it turns out, it doesn't matter in this case because 2295 is *way* beyond what we would expect from either of these distributions: the P-value is approximately 10^{-495}. So we have extremely strong support for the alternative hypothesis against the null hypothesis.

So in this example, I found extremely strong statistical support for very tiny differences in the GC content between the human X and Y chromosomes. I think this is a good example of the kinds of results that are easy to obtain with very large genomic data, but need to be interpreted with caution. Although we can come up with lots of hypotheses for why these tiny differences have biological significance, before we do that, I think it's important to remember that every statistical hypothesis test has assumptions. Let's revisit out assumptions: the null distribution of the LRT assumed that the sample size was approaching infinity. Well, actually we had millions of datapoints, so that assumption seems ok. What about the use of the multinomial distribution in the first place? We assumed that the individual letters in the genome were i.i.d. In fact, we know that assumption is almost certainly wrong: GC isocores, CpG biased mutations, repeat elements, homopolymeric runs all make the positions in the genome very highly correlated. In fact, a quick calculation shows that on the Y chromosome, there are nearly 50% more As when the preceding nucleotide is an A than when the preceding nucleotide is a T, even though overall A and T both appear very close to 30% of the time.

Thus, the highly significant (but tiny) differences that we see between the X and Y chromosomes could be biological, but they could also result from an unrealistic assumption we made when we designed the statistical test. As datasets become larger, we have more and more power to detect interesting biological effects, but we also have more power to obtain highly significant results that are due to unrealistic assumptions. Perhaps the best way to check if your statistical test is behaving as you hope is to include a biological negative control. In this example, we could perform the same test for other chromosome pairs and see if we find the same differences. Even better, we could chop up the X and Y chromosomes into large chunks and glue these back together to create random fake chromosomes. If we still find statistical differences between these, we can conclude that the differences we are seeing are artifacts of our statistical test, and not real differences between chromosomes. On the other hand, if our fake chromosomes don't show any effects, we can begin to believe that there really are differences between the X and Y chromosomes.

EXERCISES

1. What is the most probable value under a univariate Gaussian distribution? What is its probability?

2. Use the joint probability rule to argue that a multivariate Gaussian with diagonal covariance is nothing but the product of univariate Gaussians.

3. Show that the average is also the MLE for the parameter of the Poisson distribution. Explain why this is consistent with what I said about the average of the Gaussian distribution in Chapter 1.

4. Fill in the components of the vectors and matrices for the part of the multivariate Gaussian distribution:

$$\frac{1}{2}(\mu - X_i)^T \Sigma^{-1}(\mu - X_i) = [?\quad ?\quad \cdots] \begin{bmatrix} ? & ? & \\ ? & ? & \\ & & \ddots \end{bmatrix} \begin{bmatrix} ? \\ ? \\ \vdots \end{bmatrix}$$

5. Derive the MLE for the covariance matrix of the multivariate Gaussian (use the matrix calculus tricks I mentioned in the text).

6. Why did we need Lagrange multipliers for the multinomial MLEs, but not for the Guassian MLEs?

7. Notice that I left out the terms involving factorials from the multinomial distribution when I calculated the LRT. Show/explain why these terms won't end up contributing to the statistic.

REFERENCES AND FURTHER READING

Mardia KV, Kent JT, Bibby JM. (1976). *Multivariate Statistics*. London U.K: Academic Press.

Pritchard JK, Stephens M, Donnelly P. (2000). Inference of population structure using multilocus genotype data. *Genetics* 155(2):945–959.

II

Clustering

Identifying groups is one of the major tasks in biology. For example, groups of organisms define species, groups of patients define diseases, and groups of genes define pathways. In modern molecular biology, we might obtain datasets where we don't know the groups in advance: Identifying a new group might be a major discovery. Even when we do know what the groups are, identifying new members of a group is usually interesting.

Consider protein complexes, which are groups of proteins that work together to perform functions in the cell. Examples include key molecular machines, such as the ribosome and proteasome. From the perspective of the individual protein complex member proteins, membership in a complex means that they have many physical interactions with other proteins in the complex relative to their interactions with proteins that are not members of the complex. Systematically identifying new protein complexes in biological interaction networks is a clustering problem—with high-throughput protein interaction data and appropriate clustering methods, new protein complexes have been identified automatically (Bader and Hogue 2003). Similarly, genes in the same pathway, although they might not all interact directly, might tend to show similar expression patterns, or at least be more similar in gene expression than genes that are not in the same pathway. Identifying these groups of genes is another example of a clustering problem (Eisen et al. 1998).

The general idea of clustering is to take some observations from a pool and find groups or "clusters" within these observations. A simple statistical interpretation of this process is to imagine that there is really more than one kind of observation in the pool, but that we don't know that in advance. In the case of protein complexes, these different kinds of observations would be the different complexes, such as ribosome, proteasome, etc. Although we sample randomly from the pool, we actually alternate between observations of different kinds. Clustering is one of the major data analysis techniques used in molecular biology, and there are mature software packages such as Genecluster 3.0 and Cytoscape that allow cluster analysis, visualization, and interactive exploration of large datasets. I will use these to show examples of clustering in this section.

CHAPTER 5

Distance-Based Clustering

P ERHAPS THE MOST INTUITIVE way to group observations into clusters is to compare each data point to each other and put the data points that are closest together into the same cluster. For example, if the observations are 5.1, 5.3, 7.8, 7.9, 6.3, we probably want to put 5.1 and 5.3 being only 0.2 apart, in the same cluster, and 7.8 and 7.9 in the same cluster, as they are only 0.1 apart. We then have to decide what to do with 6.3: We could decide that it belongs in its own cluster or put it in one of the other two clusters. If we decide to put it in one of the other clusters, we have to decide which one to put it in: It's 1.0 and 1.2 away from the observations in the first cluster and 1.5 and 1.6 away from the observations in the second cluster. This simple example illustrates the major decisions we have to make in a distance-based clustering strategy. And, of course, we're not actually going to do clustering by looking at the individual observations and thinking about it: We are going to decide a set of rules or "algorithm" and have a computer automatically perform the procedure.

5.1 MULTIVARIATE DISTANCES FOR CLUSTERING

One important issue that is not obvious in this simple example is the choice of the distance used to compare the observations. The notion of the distance between data points is a key idea in this chapter, as all of the clustering methods we'll discuss can be used with a variety of distances. In the example mentioned earlier, it makes sense to say that 5.1 and 4.9 are

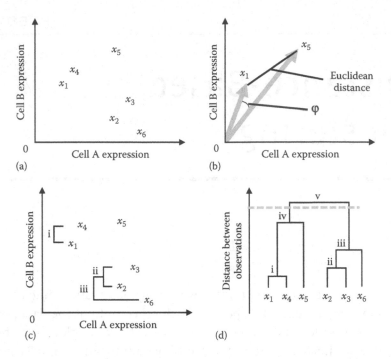

FIGURE 5.1 Grouping genes based on expression in two cell types. (a) Illustrates expression data represented as points in two dimensions. (b) Illustrates the differences between absolute distance, and the angle between observations. (c) and (d) show hierarchical clustering of gene expression data. (i–v) Indicate the steps in the iterative process that builds the hierarchical tree. The first two steps merge pairs of similar observations. The third step merges a cluster of two observations with another observation. Based on the hierarchical tree, it looks like there are two clusters in the data, and we could choose a distance cutoff (dashed line) to define the two clusters.

0.2 apart. However, in a typical genomics problem, we are doing clustering to find groups in high-dimensional data: We need to know how far apart the observations are in the high-dimensional space. Figure 5.1a illustrates gene expression observations in two dimensions.

In a typical gene-expression clustering problem, we might want to find clusters of genes that show similar expression patterns. In the case of the ImmGen data, there are thousands of genes, each of whose expression level has been measured in hundreds of cell types. To cluster these genes (or cell types) using a distance-based approach, we first need to decide on the definition of "distance" we'll use to compare the observations. Multivariate distances are most easily understood if the data are thought of as vectors in the high-dimensional space. For example, to cluster genes in the ImmGen data, each cell type would represent a different dimension, and each gene's expression levels would be a vector of length equal to the number of cell types. This is conveniently represented as a vector $X_g = (X_{g1}, X_{g2}, \ldots, X_{gn})$, where X_g is the expression vector for the gene g, for each of the n different cell types. We can then define a "distance" between two genes, g and h as some comparison of their two vectors, X_g and X_h. A typical distance that is used in machine learning applications is the "Euclidean distance," which is the sum of the squared distance between the components of two vectors.

$$d(X_g, X_h) = \sqrt{\sum_{i=1}^{n} (X_{gi} - X_{hi})^2} = \sqrt{(X_g - X_h)^T (X_g - X_h)} = \|X_g - X_h\|$$

The second last equation shows the definition using vector notation, and the last equality uses special $\|x\| = \sum_{i=1}^{d} x_i^2$ notation that indicates the length or "norm" of a vector. The distance is called Euclidean because in two or three dimensions, this is the natural definition of the distance between two points. Notice that in the Euclidean distance, each of the n-dimensions (in this case, cell types) is equally weighted.

In general, one might want to consider weighting the dimensions because they are not always independent. For example, if two cell types have very similar expression patterns *over all*, we would not want to cluster genes based on similarity of expression patterns in those cell types. The correlation between dimensions becomes particularly problematic if we are trying to cluster observations where there are many correlated dimensions and only a small number of other dimensions that have a lot of information. The large number of correlated dimensions will tend to dominate the Euclidean distance and we will not identify well-delineated clusters. This leads to the idea that distances should be weighted based on how correlated the dimensions are. An example of a popular distance that does this is the so-called Malhalanobis distance:

$$d(X_g, X_h) = \sqrt{(X_g - X_h)^T S^{-1} (X_g - X_h)}$$

which weights the dimensions according to the inverse of the covariance matrix (S^{-1}). In principle, this distance does exactly what we might want: It downweights the dimensions proportional to the inverse of their correlation. A major issue with using this distance for high-dimensional data is that you have to estimate the covariance matrix: If you have 200 cell types, the covariance is a (symmetric) 200×200 matrix where each element needs to be estimated (20,100 parameters), so you will need a lot of data!

In practice, neither Euclidean distance nor Malhalanobis distance works very well for clustering gene expression data because they consider the "absolute" distance between the vectors. Usually, we are not interested in genes with similar expression levels, but rather genes with similar "patterns." Therefore, we would like distances that don't depend on the length of the vectors. An example of a widely used distance for clustering gene expression is the so-called correlation distance that measures the similarity of the patterns.

$$d(X_g, X_h) = 1 - r(X_g, X_h) = 1 - \frac{\sum_{i=1}^{n} (X_{gi} - m(X_g))(X_{hi} - m(X_h))}{s(X_g)s(X_h)}$$

where $r(X_g, X_h)$ is Pearson's correlation between vectors X_g and X_h (which we will see again in Chapter 7), and I have used $s(X)$ to indicate the standard deviation and $m(X)$ to indicate the average (or mean) of the observation vectors. We put the "1 –" in front because the correlation is 1 for very similar vectors and −1 for vectors with opposite patterns. So this distance is 0 if vectors have identical patterns and 2 for vectors that are opposite. Notice that the correlation is "normalized" by both the mean and the standard deviation, so it is insensitive to the location and the length of the vectors: It is related (in a complicated way) to the angle between the two vectors, rather than a geometric "distance" between them. Following this idea, perhaps a more geometrically intuitive distance is the so-called cosine-distance (Eisen et al. 1998), which is more obviously related to the cosine of the angle between the vectors in the high-dimensional space:

$$d(X_g, X_h) = 1 - \cos\varphi(X_g, X_h) = 1 - \frac{X_g^T X_h}{\|X_g\|\|X_h\|}$$

where I have used the $\|x\|$ notation to indicate the norm of the observation vectors (as in the formula for Euclidean distance). Once again we put in the "1 –" because the cosine of 0° (two vectors point at an identical angle) is 1 and that of 180° (the two vectors point in opposite directions) is −1.

So once again, this distance goes between 0 and 2. Although this formula looks quite different to the one I gave for the correlation, it is actually very similar: It is a so-called "uncentered" correlation where the mean (or expectation) has not been subtracted from the observations. So this distance does depend on the location of the vectors in the high-dimensional space, but is still independent of their lengths. Figure 5.1b illustrates an example where the correlation or cosine distances would be very small, but the Euclidean or Malhalanobis distances would be large. It's important to note that although the correlation and cosine distance only considers the patterns and not the magnitude of the observations, they still treat each dimension equally and will not usually produce good clusters in practice if the data contain many highly correlated dimensions. In order to account for the possibility of correlated dimensions, we can define a weighted cosine distance as follows:

$$d(X_g, X_h|w) = 1 - \frac{X_g^T (Iw) X_h}{Tr(Iw)\|X_g\|\|X_h\|}$$

where $w = (w_1, w_2, ..., w_n)$ specifies a weight for each dimension, and I have used the identity matrix I, and some awkward notation, $Tr(Iw)$ to show that the distance must be normalized by the sum of the weights. In gene cluster 3.0 (de Hoon et al. 2004), these weights are defined heuristically, based on the intuition that genes that are similar to many other genes should be downweighted relative to genes that have few neighbors in the high dimensional space. Usually, these weights are important to getting good results on large datasets.

5.2 AGGLOMERATIVE CLUSTERING

Once the distance metric is chosen, there are still many different possible clustering strategies that can be used. I'll first describe the one most commonly used in bioinformatics and genomics applications—agglomerative hierarchical clustering. The clustering procedure starts by assigning each datapoint to its own cluster. Then, the clusters are searched for the pair with the smallest distance. These are then removed from the list of clusters and merged into a new cluster containing two datapoints, and the total number of clusters decreases by 1. The process is repeated until the entire dataset has been merged into one giant cluster. Figure 5.1c illustrates the first three steps of hierarchical clustering in the simple example. Figure 5.1d illustrates a completed hierarchical tree from this data.

There is one technical subtlety here: How to define the distances between clusters with more than one datapoint inside them? When each datapoint is assigned to its own cluster, it's clear that the distance between two clusters can be simply the distance between the two datapoints (using one of the distances defined earlier). However, once we have begun merging the clusters, in order to decide what is the closest pair in the next round, we have to be able to calculate the distance between clusters with arbitrary, different numbers of observations inside them. In practice, there are a few different ways to do this, and they differ in the interpretation of the clusters and the speed it takes to compute them. Perhaps the simplest is the so-called "single linkage" where the distance between two clusters is defined as the *smallest* distance between any pair of datapoints, where one is taken from each cluster. However, you might argue that the distance between clusters shouldn't just reflect the two closest observations—those might not reflect the overall pattern in the cluster very well. A more popular (but a little bit harder to calculate) alternative is the so-called "average linkage" where the distance between two clusters is defined as the average distance between all pairs of datapoints where one is taken from each cluster. These distances are illustrated in Figure 5.2. When average linkage is used to do agglomerative hierarchical clustering, the procedure is referred to as Unweighted Pair Group Method with Arithmetic Mean (UPGMA), and this is by far the most common way clustering is done. As long as the clustering is done by joining individual observations into groups and then merging those groups, the process is referred to as "agglomerative." If, on the other hand, the clustering approach starts with the entire pool and then tries to cut the dataset into successively smaller and smaller groups, it is known as divisive.

At first, the whole agglomerative clustering procedure might seem a bit strange: The goal was to group the data into clusters, but we ended merging all the data into one giant cluster. However, as we do the merging, we keep track of the order that each point gets merged and the distance separating the two clusters or datapoints that were merged. This ordering defines an "hierarchy" that relates every observation in the sample to every other. We can then define "groups" by choosing a distance cutoff on how large a distance points within the same cluster can have. In complex data sets, we often don't know how many clusters we're looking for, and we don't know what this distance cutoff should be (or even if there will be one distance cutoff for the whole hierarchy). But the hierarchical clustering can still be used to define clusters in the data even in an ad hoc way

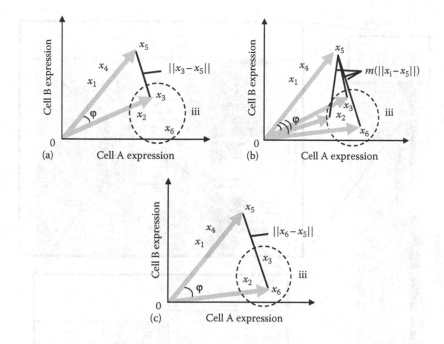

FIGURE 5.2 Distances between a cluster and a datapoint. Three observations (X_2, X_3, X_6) have been merged into a cluster (iii) and we now calculate the distance of a datapoint (X_5) from the cluster to decide whether it should be added or not. In single linkage (a), we use the distance to the closest point; in complete linkage (c), we use the distance to the furthest point; while for average linkage (b), we calculate the average of the pairwise distances between the points in the two clusters. I have indicated the average using $m()$, and used i to index the points in cluster iii. Gray arrows indicate the vectors between which we are calculating distances. The Euclidean distance is indicated as a black line, while the angle between vectors is indicated by φ.

by searching the hierarchy for groups that "look good" according to some user-defined criteria.

Hierarchies are very natural structures by which to group biological data because they capture the tree-like relationships that are very common. For example, the ribosome might appear as a cluster of genes, but this cluster might have two clusters within it that represent the 60S particle and the 40S particle. Similarly, T cells might appear as a cluster of cells, but this cluster could be made up of several subtypes of T cells, such as the cytotoxic T cells, natural killer (NK) T cells, and helper T cells.

In the following example (Figure 5.3), I have hierarchically clustered both the genes and cell types: For the genes, this means finding the groups

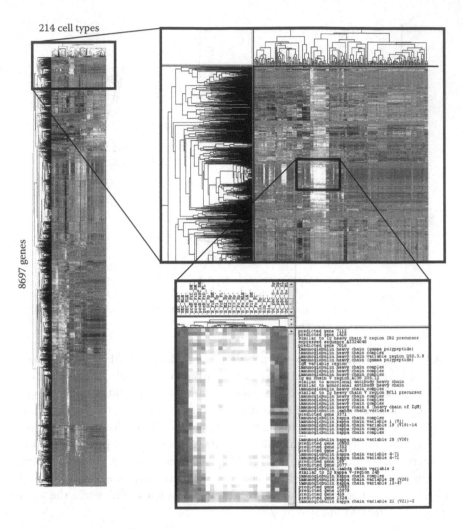

FIGURE 5.3 Hierarchical clustering of the ImmGen data using GeneCluster 3.0. Hierarchical clustering with average linkage for 8697 genes (that had the biggest relative expression differences) in 214 cell types. Both genes and cell types were weighted using default parameters. The data matrix is displayed using a heatmap where white corresponds to the highest relative expression level, gray corresponds to average expression, and black corresponds to the lowest relative expression (gene expression was log-transformed and mean log-expression level was subtracted from each gene). Left shows a representation of the entire dataset. Right top panel shows a clear "cluster" where the distances (indicated by the depth and height of the dendogram) are small between genes and cells. Right bottom panel shows that this corresponds to a group of immunoglobulin genes that are highly expressed in B cells. The clustering results are visualized using Java TreeView.

in a 214-dimensional space (each gene has a measurement in each cell type), while for the cell types, this means finding the groups in an 8697-dimensional space (each cell type has a measurement for each gene). The tree (or dendogram) created for the genes and cell types are represented beside the data. In general, if possible, it's always preferential to represent a clustering result where the primary data are still visible. Often, visualizing the primary data makes immediately apparent that clusters (or other structures in the data) are driven by technical problems: missing data, outliers, etc.

5.2 CLUSTERING DNA AND PROTEIN SEQUENCES

To illustrate the power of thinking about molecular biology observations in high-dimensional spaces, I want to now show that a nearly identical clustering procedure is used in an even more "classical" problem in evolutionary genetics and molecular evolution. Here, the goal is to hierarchically organize sequences to form a tree. In this case, the hierarchical structure reflects the shared ancestry between the sequences. Although the task of organizing sequences hierarchically is widely referred to as "phylogenetic reconstruction," in a sense, reconstructing a phylogenetic tree is a way of grouping similar sequences together—the more similar the sequences in the group, the more recently they shared a common ancestor. To see the mathematical relationship between this problem and clustering gene expression data, we first need to think about the sequences as multivariate observations that have been drawn from a pool. In a sequence of length, L, each position, j, can be thought of as a vector X_j, where the bth component of X_j is 1 if the sequence has the DNA base b at that position. For example, the DNA sequence CACGTG would be

$$X = \begin{matrix} 0 & 1 & 0 & 0 & 0 & 0 \\ 0 & 0 & 0 & 1 & 0 & 1 \\ 1 & 0 & 1 & 0 & 0 & 0 \\ 0 & 0 & 0 & 0 & 1 & 0 \end{matrix}$$

where the bases have been ordered arbitrarily from top to bottom as, A, C, G, T. A protein sequence might be represented as a $20 \times L$ matrix. To compute the distance between two (aligned) gene sequences, X_g and X_h, we could use one of the distances defined earlier. For example, we could sum up the cosine distance at each position

$$d(X_g, X_h) = \sum_{j=1}^{L} \left(1 - \frac{X_{gj}^T X_{hj}}{||X_{gj}|| ||X_{hj}||} \right) = L - X_g^T X_h$$

where to derive the simple formula on the right, I have used the fact that the norm of the vector at each position in each gene sequence is 1 (each position can only have one 1, all the others must be 0s), and the inner product between two matrices is the sum of the dot products over each of the component vectors. In sequence space, this distance metric is nothing but *the number of different nucleotide (or protein) residues* between the two vectors and is a perfectly sensible way to define distances between closely related sequences. Notice that just like with gene expression distances, this simple metric treats all of the dimensions (types of DNA or protein residues) identically. In practice, certain types of DNA (or protein) changes might be more likely than others, and ideally these should be downweighted when computing a distance.

In practice, for comparison of sequences, evolutionary geneticists and bioinformaticists have developed so-called "substitution" or "similarity" matrices that are used to define the distance between sequences. Indeed, perhaps one of the lasting contributions of the earliest bioinformatics projects were Marget Dayhoff's so-called PAM (Point Accepted Mutation) matrices (Strasser 2010). Changes between DNA (or protein) residues are not equally likely, and therefore bioinformaticists defined quantitative values related to the likelihood of each type of change to capture the idea that not all residues are equally far apart biologically. Figure 5.4 shows the BLOSUM62 substitution matrix, one of the matrices currently in wide use for protein sequence analysis. Taking into account these weights, the distance between two (aligned) sequences is then

$$d(X_1, X_2) = -\sum_{j=1}^{L} \sum_{a \in \{A,C,G,T\}} \sum_{b \in \{A,C,G,T\}} X_{1ja} M_{ab} X_{2jb} = -\sum_{j=1}^{L} X_{1j}^T M X_{2j} = -X_1^T M X_2$$

where M is a "similarity" matrix defining a score for each pair of DNA (in this case) residues, a and b. Notice that to make the similarity into a distance, I simply put a negative sign in front. In the case where $M = I$ (the identity matrix), the distance between two sequences is simply proportional to the number of different positions. This type of distance is directly analogous to the weighted distances discussed above. Once these distances have been defined, it is straightforward to find the "closest" or most similar pairs of sequences in the sample. The closest sequences are then iteratively merged as described earlier to build the bifurcating tree, and clustering of sequences can proceed just as any other data.

	A	R	N	D	C	Q	E	G	H	I	L	K	M	F	P	S	T	W	Y	V	B	Z	X	*
A	4	-1	-2	-2	0	-1	-1	0	-2	-1	-1	-1	-1	-2	-1	1	0	-3	-2	0	-2	-1	0	-4
R	-1	5	0	-2	-3	1	0	-2	0	-3	-2	2	-1	-3	-2	-1	-1	-3	-2	-3	-1	0	-1	-4
N	-2	0	6	1	-3	0	0	0	1	-3	-3	0	-2	-3	-2	1	0	-4	-2	-3	3	0	-1	-4
D	-2	-2	1	6	-3	0	2	-1	-1	-3	-4	-1	-3	-3	-1	0	-1	-4	-3	-3	4	1	-1	-4
C	0	-3	-3	-3	9	-3	-4	-3	-3	-1	-1	-3	-1	-2	-3	-1	-1	-2	-2	-1	-3	-3	-2	-4
Q	-1	1	0	0	-3	5	2	-2	0	-3	-2	1	0	-3	-1	0	-1	-2	-1	-2	0	3	-1	-4
E	-1	0	0	2	-4	2	5	-2	0	-3	-3	1	-2	-3	-1	0	-1	-3	-2	-2	1	4	-1	-4
G	0	-2	0	-1	-3	-2	-2	6	-2	-4	-4	-2	-3	-3	-2	0	-2	-2	-3	-3	-1	-2	-1	-4
H	-2	0	1	-1	-3	0	0	-2	8	-3	-3	-1	-2	-1	-2	-1	-2	-2	2	-3	0	0	-1	-4
I	-1	-3	-3	-3	-1	-3	-3	-4	-3	4	2	-3	1	0	-3	-2	-1	-3	-1	3	-3	-3	-1	-4
L	-1	-2	-3	-4	-1	-2	-3	-4	-3	2	4	-2	2	0	-3	-2	-1	-2	-1	1	-4	-3	-1	-4
K	-1	2	0	-1	-3	1	1	-2	-1	-3	-2	5	-1	-3	-1	0	-1	-3	-2	-2	0	1	-1	-4
M	-1	-1	-2	-3	-1	0	-2	-3	-2	1	2	-1	5	0	-2	-1	-1	-1	-1	1	-3	-1	-1	-4
F	-2	-3	-3	-3	-2	-3	-3	-3	-1	0	0	-3	0	6	-4	-2	-2	1	3	-1	-3	-3	-1	-4
P	-1	-2	-2	-1	-3	-1	-1	-2	-2	-3	-3	-1	-2	-4	7	-1	-1	-4	-3	-2	-2	-1	-2	-4
S	1	-1	1	0	-1	0	0	0	-1	-2	-2	0	-1	-2	-1	4	1	-3	-2	-2	0	0	0	-4
T	0	-1	0	-1	-1	-1	-1	-2	-2	-1	-1	-1	-1	-2	-1	1	5	-2	-2	0	-1	-1	0	-4
W	-3	-3	-4	-4	-2	-2	-3	-2	-2	-3	-2	-3	-1	1	-4	-3	-2	11	2	-3	-4	-3	-2	-4
Y	-2	-2	-2	-3	-2	-1	-2	-3	2	-1	-1	-2	-1	3	-3	-2	-2	2	7	-1	-3	-2	-1	-4
V	0	-3	-3	-3	-1	-2	-2	-3	-3	3	1	-2	1	-1	-2	-2	0	-3	-1	4	-3	-2	-1	-4
B	-2	-1	3	4	-3	0	1	-1	0	-3	-4	0	-3	-3	-2	0	-1	-4	-3	-3	4	1	-1	-4
Z	-1	0	0	1	-3	3	4	-2	0	-3	-3	1	-1	-3	-1	0	-1	-3	-2	-2	1	4	-1	-4
X	0	-1	-1	-1	-2	-1	-1	-1	-1	-1	-1	-1	-1	-1	-2	0	0	-2	-1	-1	-1	-1	-1	-4
*	-4	-4	-4	-4	-4	-4	-4	-4	-4	-4	-4	-4	-4	-4	-4	-4	-4	-4	-4	-4	-4	-4	-4	1

FIGURE 5.4 An example of similarity matrix used in bioinformatics, BLOSUM62 (Henikoff and Henikoff 1992). The 20 naturally occurring amino acids (and X and *, which represent ambiguous amino acids and stop codons, respectively) are represented. Higher scores mean greater similarity—entries along the diagonal of the matrix tend to be relatively large positive numbers.

Note that the agglomerative hierarchical clustering method described does not actually estimate any parameters or tell you which datapoints are in which clusters. Nor does it have an explicit objective function. Really it is just organizing the data according to the internal structure of similarity relationships. Another important issue with hierarchical clustering is that it is necessary to compute the distances between all pairs of datapoints. This means that if a dataset has thousands of observations, milllions of distances must be calculated, which, on current computers, is typically possible. However, if the dataset approaches millions of datapoints, the number of pairwise distance calculations needed becomes too large even for very powerful computers.

5.4 IS THE CLUSTERING RIGHT?

UPGMA and other hierarchical clustering methods are ubiquitously used in molecular biology for clustering genome-scale datasets of many kinds: gene expression, genetic interactions, gene presence and absence, etc.

Perhaps the biggest conceptual problem molecular biologists encounter with agglomerative clustering is that it doesn't tell you whether you have the "right" or "best" or even "good" clusters. Let's start with the question of how to decide if you have good clusters. For example, let's say you don't know whether to choose average linkage or single linkage clustering—so you try both. Which one worked better?

Ideally, you would be able to decide this based on some external information. For example, if you knew for a few of the data points that they belonged in the same or different groups, you could then ask for each set of clustering parameters, whether these datapoints were clustered correctly. In molecular biology, this is very often possible: If one is clustering genes based on some genomic measurements, a sensible way to evaluate the quality of the clustering is to perform gene set enrichment analysis, and choose the clusters that give (on average, say) the most significant association with other data.

However, what about a situation where you don't know anything about your data in advance? In fact, there are sensible metrics that have been proposed over the years for summarizing how well a dataset is organized into clusters, even without any external information. I will discuss just one example here, known as the silhouette (Rousseeuw 1987). The silhouette compares how close a datapoint is to the cluster that it has been assigned to, relative to how close it is to the closest other cluster (that it has

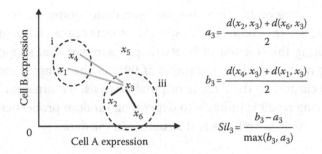

$$a_3 = \frac{d(x_2, x_3) + d(x_6, x_3)}{2}$$

$$b_3 = \frac{d(x_4, x_3) + d(x_1, x_3)}{2}$$

$$Sil_3 = \frac{b_3 - a_3}{\max(b_3, a_3)}$$

FIGURE 5.5 The silhouette for a single data point. In the left panel, two clusters are indicated (i and iii). The distances between a single data point (X_3) to the datapoints in the cluster it is assigned to are indicated by black lines, while the distances to the datapoints in the nearest cluster are indicated by gray lines. The average of these distances (a_3 and b_3, respectively) is shown on the right. The silhouette is the ratio of the difference of these averages to their maximum.

not been assigned to). Figure 5.5 illustrates the idea. The average silhouette over a whole dataset for a given clustering method measures how good the clustering is overall.

Possibly more important (and more difficult) than asking whether the clustering is good is whether a datapoint belongs to one cluster (as opposed to another) or whether there are two groups in the data or just one. Clustering is meant to be exploratory data analysis and therefore doesn't really have a strong framework for hypothesis testing. However, these questions can become very important if you are using clustering to discover biological structure in the data. For example, if you are clustering tumor samples based on gene expression, you might have grouped the samples into two clusters, and then you want to know whether really there is statistical evidence for the two subtypes. The distribution of the silhouette for the datapoints in each cluster can help decide if the cluster is really justified. For example, if the average silhouette of the points in a cluster is very small, this means that the data in the cluster are just as close to their cluster as to the neighboring cluster. This indicates that there wasn't really a need for two separate clusters.

It's also possible to test whether the pattern you found in the data using clustering is robustly supported by bootstrapping. The idea of bootstrapping is to randomly resample from the dataset (as if it was the pool that observations were originally drawn from) and then rerun the clustering. Typically, bootstrapping is done by leaving out some of the dimensions or some of the datapoints. After this has been done thousands of times, we can summarize the confidence in specific aspects of the clustering results by calculating the fraction of bootstrapped samples that supported the clustering we observed. For example, if 99% of the samples showed the same two clusters in the data as original analysis, we can conclude that our clustering result is unlikely to depend on random properties of input dataset, but probably reflects real structure in the data.

TECHNIQUES TO EVALUATE CLUSTERING RESULTS
- Comparison with external data: If you know the true clusters for some of your data, compare your clustering result with that.
- Even if you don't know the true clusters, you can try to maximize the statistical association of your clusters with some external data (such as GO annotations if you are clustering genes).
- Statistical measures of purity or homogeneity of the cluster such as silhouette: Use a metric that lets you quantitatively compare the clustering results under different parameter settings.
- Bootstrap your results: Bootstrapping gives you confidence that the data uniformly support the clustering conclusions.

5.5 *K*-MEANS CLUSTERING

K-means clustering is really the first technique that we'll see in this book where the computer will actually "learn" something. In some sense, it is the most basic "machine learning method." Note that *K*-means is sometimes called "*c*-means" in the statistical literature, but in the machine learning world, it's definitely *K*-means. The idea of *K*-means is to take seriously the idea that there are underlying groups in the data, and that each of these clusters (exactly *K* of these clusters) can be summarized by its mean. So *K*-means aims to assign each datapoint to one of *K* clusters, based on the similarity of the data to the mean of the cluster.

In particular, *K*-means says: Assign each datapoint to the cluster to which it is most similar. Formally, if X is a series of data vectors, $X_1, X_2, \ldots, X_n$ we can write this as the *K*-means cluster assignment rule:

$$Z_{ik} = 1 \quad \text{if } \arg\min_c [d(m_c, X_i)] = k$$

$$Z_{ik} = 0 \quad \text{otherwise}$$

Here, I have used some special notation. First, I have used $d(m, X)$ to represent the distance between the mean for the cluster and the observation vector, X, but in traditional K-means clustering, the distance is always taken to be the Euclidean distance. K-means, with other distances, has other names (e.g., spherical K-means refers to K-means with the cosine distance described). Most important, however, is the variable Z that indicates which cluster observation i is assigned to. We define this variable to take on the values 1 or 0 (think of these as true or false). The rule says that Z is 1 if the mean, m, for cluster k is closer to X_i than any other of the other means. Make sure you understand what this "indicator variable" Z means—it is a key notational trick that will allow us to write some very fancy formulas.

But what are these "mean" patterns of the clusters? They are nothing more than the average patterns of all the data points assigned to them. We can write this as the K-means mean estimate:

$$m_k = \frac{1}{\sum_{i=1}^{n} Z_{ik}} \sum_{i=1}^{n} Z_{ik} X_i$$

Notice how cleverly I used the indicator variable Z in this formula. Although I said that this is nothing more than the average, you can see that the formula is more complicated than a normal average. First, I used the indicator variable to include (or exclude) the observations X_i that are (not) assigned to cluster k, by multiplying them by the indicator. Because the indicator is 1 when the observation is assigned to the cluster, those values are included in the sum, and because the indicator is 0 for the observations that are not assigned to the cluster, they are multiplied by 0 and therefore do not contribute to the sum. Instead of a normal average, where I would divide by the total number of observations, now I divided by the sum of the indicator variable: This adds 1 for each of the datapoints in the cluster, and 0 for any data that are not assigned to the cluster. Using this indicator variable, I can write a single formula to take the averages of all the clusters without having to define variables for the numbers of genes in each cluster, etc. It might all seem a bit strange, but I hope it will become clear that indicator variables are actually incredibly powerful mathematical tricks.

OK, enough about the indicator variable already! We're supposed to be doing clustering, and how can this K-means idea work: If we don't know the means, m, to start with, how do we ever assign data points to clusters (according to the cluster assignment rule)? And if we can't assign data-points to clusters, then how do we ever calculate the means (using the means estimate)?

In fact, there are several answers to these questions, but the most common answer is: Start with *random* assignments of datapoints to clusters. Although this will produce a very bad set of means, we can then alternate between reassigning all of the datapoints to clusters and then recalculating the means for each cluster (Figure 5.6). This type of procedure is known as an "iterative" procedure, and nearly all of the machine learning methods

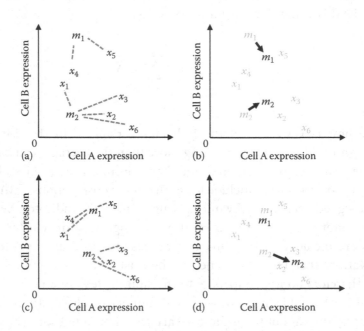

FIGURE 5.6 How the K-means algorithm "learns." (a) Illustrates the initial (wrong) assignment (indicated by dashed gray lines) of observations to randomly chosen cluster means, m. (b) Shows that when the means are recalculated based on the observations, they will be pulled towards the majority of their datapoints. This way, datapoints that are far away from the cluster will be "left behind". (c) Shows the reassignment (dashed gray lines) of observations to the means in the next iteration. (d) Finally, once the "right" datapoints are assigned, the means end up very close to the centers of the clusters.

we'll consider will lead to these types of iterative procedures. The main idea is that the computer starts of knowing nothing—it takes a random guess at the parameters (in this case, the means of the clusters). But then, if the rules for updating the parameters are sensible, the estimates of the parameters will get better and better with each iteration—the machine will *learn*.

If we consider the objective function of K-means to be the sum of the distances between the datapoints and the mean of the cluster they've been assigned to, we can formalize the optimization problem as trying to minimize the function of the means, m.

$$f(m_1, m_2, \ldots, m_k) = \sum_{c=1}^{K} \sum_{i=1}^{n} d(X_i, m_c) Z_{ic}$$

In this formula, we have to sum over all the K clusters, so I've introduced c to index them. Notice again the clever use of the indicator variable Z to multiply all of the distances between the data and the clusters that they are not assigned to by 0, so they don't contribute to the objective function.

It's important to think about how complicated this function really is: Each of the K-means actually is a vector of length equal to the dimensionality of the data. In the case of the ImmGen data, this is a vector of length 200. Since we have K of these means, we might have $K*200$ parameters to find. Because the parameters are actually determined by the assignments of each datapoint to the clusters, really there are on the order of K^n possible combinations. This is an incredibly large number even for a small number of clusters, because there are typically thousands of datapoints.

The good news is that the iterative procedure (algorithm) described is guaranteed to decrease the objective function in each step—decreasing the distance between the cluster means and the datapoints is a good thing—it means the clusters are getting tighter. However, the algorithm is not guaranteed to find the global minimum of the objective function. All that's guaranteed is that in each step, the clusters will get better. We have no way to know that we are getting the best possible clusters. In practice, the way we get around this is by trying many random starting guesses and hope that we've covered the whole space of the objective function.

To illustrate the use of the K-means algorithm on a real (although not typical) dataset, consider the expression levels of CD8 and CD4 in the immune cells in the ImmGen data. As is clear from Figure 5.7, there are four types of cells, CD8−CD4−, CD8−CD4+, CD8+CD4− and CD4+CD8+. However, this is a difficult clustering problem, first, because

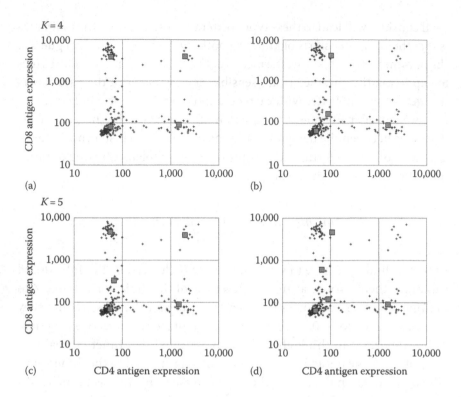

FIGURE 5.7 K-means clustering of cell-types in the ImmGen data. Each black "+" indicates one cell type, and gray squares identify the means identified by the K-means algorithm. (a) Shows convergence to the "right" clusters with $K=4$. (b) Shows the tendency of the K-means algorithm to split up the large cluster with low expression for both genes. (c) Shows that adding an extra cluster can help, but that the algorithm still splits up the big cluster for some initializations (d).

the sizes of the clusters are not equal. This means that the contribution of the large (CD4−CD8−) cluster to the objective function is much larger than the very small cluster (CD8+CD4+). In addition to the unequal sizes of the clusters, there are some cells that seem to be caught in between the two CD4− clusters. Because of this, for some random choices of starting cluster means, K-means chooses to split up the large cluster. This has the effect of incurring a large penalty for those datapoints (notice that they are consistently added to the CD8+CD4− cluster leading to a slight shift in the mean), but the algorithm is willing to do this because it is "distracted" by reducing the small penalties for the large number of datapoints in the large cluster. One possible solution is to try to just add more clusters—if we allow K-means to have an extra mean or two, then the algorithm can still split up the large cluster (−−), and find the smaller ++ cluster as well. This is a reasonable strategy if you really want to use K-means to find the small clusters. However, it is still not guaranteed to work as there are still many more datapoints in the big (−−) cluster, and its shape is not well-matched to the symmetrical clusters assumed by K-means.

"LEARNING SIGNAL" IN THE K-MEANS ALGORITHM

One very powerful way to think about any iterative machine learning algorithm is the idea of the "learning signal." The idea here is that if the iterative procedure is to be effective, the values of the parameters in the next iteration should be a little "better" than they currently are. In order for the parameters to improve, there has to be some sort of push or pull in the right direction. A good example of a learning signal might be some measure of the "error"— larger error might be a signal that we are going in the wrong direction, and a reduction in the error might be a signal that we are going in the right direction. In the case of the K-means algorithm, the parameters we update are the means of each cluster, and the error that we would like to minimize is the distance between each observation and its corresponding mean. Each observation therefore "pulls" the mean of the cluster toward it. Consider an iteration where we will add one observation, q, to the kth cluster. If we let the number of observations assigned to the kth cluster be $n_k = \sum_{i=1}^{n} Z_{ik}$, the new mean for the kth cluster once the new observation had been added is exactly

$$(m_k)_{new} = \frac{X_q + \sum_{i=1}^{n} Z_{ik}X_i}{1+n_k} = \frac{X_q}{1+n_k} + \frac{n_k \sum_{i=1}^{n} Z_{ik}X_i}{(1+n_k)n_k} = \frac{X_q}{1+n_k} + \frac{n_k}{1+n_k}m_k$$

where I have added the qth observation to all the others previously in the cluster and increased the denominator (the number of points in the average) by one. So the change in the mean due to the additional observation being assigned to the cluster is

$$(m_k)_{new} - m_k = \frac{X_q}{1+n_k} + \frac{n_k}{1+n_k}m_k - m_k = \frac{X_q}{1+n_k} - \frac{m_k}{1+n_k} = \frac{X_q - m_k}{1+n_k}$$

which is a vector in the direction of the difference between the new observation and the mean. Similarly for an observation being removed from a cluster, we have

$$(m_k)_{new} = \frac{\sum_{i=1}^{n} Z_{ik}X_i - X_q}{n_k - 1} = \frac{n_k \sum_{i=1}^{n} Z_{ik}X_i}{(n_k - 1)n_k} - \frac{X_q}{n_k - 1} = \frac{n_k}{n_k - 1}m_k - \frac{X_q}{n_k - 1}$$

and the change is

$$(m_k)_{new} - m_k = \frac{n_k}{n_k - 1} m_k - \frac{X_q}{n_k - 1} - m_k = \frac{m_k}{n_k - 1} - \frac{X_q}{n_k - 1} = -\frac{X_q - m_k}{n_k - 1}$$

which is also a vector in the direction of the difference between the observation and the mean. We can add up all of these effects, and write the change in mean in one iteration using clever indicator variable notation:

$$(m_k)_{new} - m_k = \sum_{i=1}^{n} \frac{(Z_{ik})_{new} - Z_{ik}}{n_k + (Z_{ik})_{new} - Z_{ik}} (X_i - m_k)$$

$$= \sum_{i=1}^{n} \frac{(Z_{ik})_{new} - Z_{ik}}{\sum_{i=1}^{n} Z_{ik} + (Z_{ik})_{new} - Z_{ik}} (X_i - m_k)$$

If you can understand this formula, you are a master of indicator variables and are ready to tackle machine learning literature. Notice how only the observations whose cluster assignments change will be included in the sum (all others will be multiplied by 0).

Anyway, the point of this formula was to show that the change in mean in the next iteration will be a weighted sum of the observations that have been added and removed from the cluster, along the direction of the vector between the observation and the mean. This can be thought of as the "learning signal": The new observations that have arrived have pulled the mean in their direction, and the observations that have left the cluster have given up pulling (or pushed away) the cluster mean. The observations are telling the mean that it's not quite in the right place to have the smallest error.

5.6 SO WHAT IS LEARNING ANYWAY?

So far, I've been a bit cavalier in using the term learning, and I've said that *K*-means is really the first algorithm we've seen where the computer will really learn something. Then I described an iterative procedure to infer the parameters that describe clusters in the data. I think it's important that you've now seen the kind of thing that machine learning really is before we get too carried away in analogy.

Despite these modest beginnings, I hope that you're starting to get some feeling for what we mean by learning: A simplified representation of the observations that the algorithm will automatically develop. When we say that "learning" has happened, we mean that a description of the data has now been stored in a previously naïve or empty space. This type of learning can be quantified using ideas from information theory: If we can measure how much information was in the model to begin with, and then run our learning algorithm (i.e., infer some parameter values) and measure how much information we have after, we can say the amount that the machine learned is the difference between the two. How information is actually measured is beyond the scope of this book, but it turns out to be simpler than one would imagine.

Granted, K-means (and most of the models that we will consider in this book) actually contains relatively little information about the data—a 100 parameters or so—so the computer doesn't actually learn very much. In fact, the amount that can be learned is limited on the one hand by how much information there really is in the data (and how much of it is really noise), and on the other by the complexity of the model (how many parameters are available) and the effectiveness of the learning algorithm that is used to train it (do we reliably extract the information in the data and store in the parameters). A key issue that arises in machine learning is that we tend to overestimate how much the computer has learned because it's hard to know when the model is learning about the real information in the data, as opposed to learning something about the noise in this particular dataset. This relates to the issues of overfitting and generalizability, and we will return to it throughout the book.

These are important (although rarely appreciated) considerations for any modeling endeavor. For example, learning even a single real number (e.g., 43.9872084...) at very high precision could take a lot of accurate data. Famously, Newton's gravitational constant is actually very hard to measure. So, even though the model is extremely simple, it's hard to learn the parameter. On the other hand, a model that is parameterized by numbers that can only be 0 or 1 (yes or no questions) can only describe a very limited amount of variation. However, with the right algorithm, it might be possible to learn these from much less (or less accurate) data. So as you consider machine learning algorithms, it's always important to keep in mind what is actually being learned, how complex it is, and how effective the inference procedure (or algorithm) actually is.

5.7 CHOOSING THE NUMBER OF CLUSTERS FOR K-MEANS

One important issue with K-means is that, so far, we have always assumed that K was chosen beforehand. But in fact, this is not usually the case: Typically, when we have 1000s of datapoints, we have some idea of what the number of clusters is (maybe between 5 and 50) but we don't know what the number is. This is a general problem with clustering: The more clusters there are, the easier it is to find a cluster for each datapoint that is very close to that datapoint. The objective function for K-means will always decrease if K is larger.

Although there is no totally satisfying solution to this problem, the measures for quantitatively summarizing clusters that I introduced can be used when trying to choose the number of clusters. In practice, however, because K-means gives a different result for each initialization, the average silhouette will be different for each run of the clustering algorithm. Furthermore, the clustering that produces the minimum average silhouette might not correspond to the biologically intuitive clustering result. For example, in the case of CD4 and CD8 expression examples, the silhouette as a function of cluster number is shown in Figure 5.8. The average silhouette reaches a maximum at $K = 3$, which is missing the small CD4+ CD8+ cluster. Since the missing cluster is small, missing it doesn't have much effect on the silhouette of the entire data—the same reason that it's hard for K-means to find this cluster in the first place.

Unfortunately (as far as I know), there is no simple statistical test that will tell you that you have the correct number of clusters. In Chapter 6,

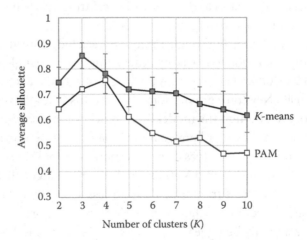

FIGURE 5.8 Choosing the number of clusters in the CD4 CD8 example using the average silhouette. For K-means (gray symbols), 100 initializations were averaged for each cluster number, and error bars represent the standard deviation of the average silhouette for each initialization. The optimal number of clusters is 3. For PAM (K-medoids, unfilled symbols), each initialization converged to the same exemplars, so there was no variation in the average silhouette. The optimal number of clusters is 4.

we'll see how to do clustering using probabilistic methods where the objective function is a likelihood. In that case, there are more principled ways of trading off the model fit to the number of parameters. In particular, we'll see that the AIC is an effective way to choose the number of clusters. Choosing the "right" number of clusters is such a difficult and pervasive issue that this alone is sometimes enough to persuade researchers to stick with hierarchical clustering, where this choice is sort of hidden. Another approach is to give up on the idea that there is a single true number of clusters and use Bayesian statistics to estimate the distribution of cluster numbers (we will return to this idea in Chapter 9). However, regardless of the solution you choose, this is something that will have to be discussed in any cluster analysis.

5.8 K-MEDOIDS AND EXEMPLAR-BASED CLUSTERING

In K-means, we explicitly defined the average pattern of all the datapoints assigned to the cluster to be the pattern associated with each cluster. However, as with the different types of linkages for hierarchical clustering, there are K-means-like clustering procedures that use other ways to define the pattern for each cluster. Perhaps the most effective of these are methods where one datapoint is chosen to be the representative or "exemplar" of that cluster. For example, one might choose the datapoint that has on average the smallest distance to all the other datapoints in the cluster as the "exemplar"of the cluster. Alternatively, one might choose exemplars that minimize the total distance between each datapoint and their closest exemplars. If this rule is chosen, the clustering is referred to as "k-medoids" and the algorithm usually used is called "partitioning around medoids" or PAM for short (no relationship with Dayhoff's PAM matrices mentioned earlier).

In this case, the objective function is

$$f(K) = \sum_{c=1}^{K} \sum_{i=1}^{n} d(X_i, X_c) Z_{ic}$$

where $d(X_i, X_c)$ represents the distance between the ith observation and the "exemplar" or "medoid" for the cth cluster. Like in traditional K-means clustering, the distance for PAM is usually taken to be the Euclidean distance. Notice the difference between this objective function and K-means: We no longer have "parameters" that represent the "means" for each cluster. Instead, we are simply trying to choose the K exemplars and the assignments of each gene to each cluster, which I again summarize using the indicator variable, Z.

Exemplar-based clustering has some important advantages. First, the distances between the clusters and all the other datapoints certainly don't need to be recalculated during the optimization: they are simply the distances between the datapoints and the exemplar. Perhaps, more importantly, there are no parameters that need to be "learned" by these algorithms (e.g., no means to be estimated in each iteration). In some sense, because exemplar-based clustering is only trying to make a yes-or-no decision about whether a data point should be the exemplar or not, it should be easier to learn this from noisy data than a vector of real numbers. Indeed, exemplar-based algorithms can often work better in practice, and K-medoids is a favorite method of statisticians. In the CD4 CD8 expression example, PAM always converged to the same answer, and for $K = 4$, always identified the small cluster of CD4+ and CD8+ cells that K-means often missed (Figure 5.8). Also, PAM shows maximum average silhouette at $K = 4$, which by visual inspection seems to be the right answer for this dataset.

Another interesting advantage is that exemplar-based methods imply a simple strategy for interpretation of the clusters: From a dataset of n observations, the user is left with, say, K exemplars—actual data points that can be examined. Conveniently, these K datapoints represent the major patterns in the data. In the CD4 and CD8 expression data, the exemplar cell identified for the small CD4+ and CD8+ cluster is called a CD8+ dendritic cell, consistent with the interpretation of that cluster.

An important downside to exemplar-based methods is that the algorithms for optimizing their objective functions can be more complicated and might be more prone to failing to reach the global maximum. Developing advanced algorithms for exemplar-based clustering is currently an important research area in machine learning (Mézard 2007).

5.9 GRAPH-BASED CLUSTERING: "DISTANCES" VERSUS "INTERACTIONS" OR "CONNECTIONS"

The clustering methods and examples that we've discussed so far all assume that we can define a distance measure in a high-dimensional space that is quantitatively meaningful (e.g., 5.1 is less than 5.3, which is less than 5.5). However, often we're in a situation where the quantitative value of distance we calculate doesn't really reflect anything interesting. In that case, we might simply want to know if two datapoints are "close" or "far." In this case, pairwise distances are reduced to simply 1 (for close) and 0 (for far). We can think of these as "interactions" or "connections" between datapoints rather than "distances," but any distance can be turned into an interaction or connection simply by placing a threshold on the distance. Once we begin to think of datapoints as connected, it's natural to think of the data forming a network or "graph" (the mathematical term for what biologists call a network). Figure 5.9 shows how pairwise distances can be turned into interactions, and then into a network or graph. Special clustering and data visualization methods (that I won't cover in this book, but are implemented in popular bioinformatics softwares such as Cytoscape,

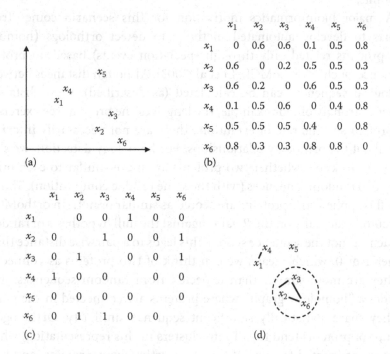

	x_1	x_2	x_3	x_4	x_5	x_6
x_1	0	0.6	0.6	0.1	0.5	0.8
x_2	0.6	0	0.2	0.5	0.5	0.3
x_3	0.6	0.2	0	0.6	0.5	0.3
x_4	0.1	0.5	0.6	0	0.4	0.8
x_5	0.5	0.5	0.5	0.4	0	0.8
x_6	0.8	0.3	0.3	0.8	0.8	0

(a) (b)

	x_1	x_2	x_3	x_4	x_5	x_6
x_1		0	0	1	0	0
x_2	0		1	0	0	1
x_3	0	1		0	0	1
x_4	1	0	0		0	0
x_5	0	0	0	0		0
x_6	0	1	1	0	0	

(c) (d)

FIGURE 5.9 Distances can be converted into interactions or connections. (a) Data in a two-dimensional space. (b) The matrix of pairwise distances between the datapoints. By applying a threshold (in this case, 0.3) to define "close" or interacting datapoints, which convert the distance matrix to a matrix of 0 and 1, shown in (c). Distances along the diagonal are omitted by convention, but they would all be 1 because each datapoint is always similar to itself. (d) Data can then be visualized by drawing lines connecting interacting datapoints. Graph clustering can then be applied to find the densely connected subgraphs within the interaction network, which is indicated as a dashed circle.

Shannon et al. 2003) can be applied to these graphs to find the highly connected regions. An advantage of the representation as binary interactions is that only the connections matter (the 1s in the matrix), and we don't need to bother keeping track of the datapoints that are not connected (the 0s in the matrix). This has an intuitive elegance: Clustering is about finding groups that are close together, so we shouldn't waste time analyzing datapoints that are far apart—we know these won't be in the same cluster. Indeed, because of this simplification, graph-based clustering algorithms can be much faster than UPGMA (which considers all possible datapoints no matter how far they are from those already in the cluster). Graph-based clustering algorithms can find thousands of clusters in datasets of millions of points.

A major bioinformatics motivation for this scenario comes from efforts to develop automated methods to detect orthologs (homologus proteins related only through speciation events) based on protein sequence (such as orthoMCL, Li et al. 2003). Although distances between biological sequences can be calculated (as described), these distances depend strongly on, for example, the length of the protein (see exercises) or protein-specific rate of evolution. These are not necessarily informative about the orthology relationships. For homology detection, we simply want to know whether two proteins are more similar to each other than two random sequences (with the same residue composition). Thus, a cutoff is applied, and proteins are scored as similar or not. (In orthoMCL, the cutoff is actually on the P-value against the null hypothesis of random sequences, not the sequence score.) This leads to a pairwise distance that's either 1 or 0, which means we can think of two proteins as connected if they are more similar than expected from random sequences. This defines a "homology graph" where proteins are connected in the graph if they share statistically significant sequence similarity. Orthologous groups of proteins tend to fall into clusters in this representation, where they are connected to most of the other orthologous proteins, and have few connections to other sequences. OrthoMCL uses a graph-clustering approach (called MCL, Enright et al. 2002) to identify more than 100,000 clusters in a graph containing more than 1 million protein sequences and 500 million pairwise connections: This is a 1 million $\times$ 1 million matrix with 500 million entries that are 1s.

Another application of graph-based clustering is to find clusters in protein interaction networks (Bader and Hogue 2003). In this case, usually the data lends itself to representation as binary interactions either for experimental reasons (the experiment was a yeast-two-hybrid) or conceptual simplicity (we want to think of proteins as simply interacting or not, and not consider multiple cell types, conditions, etc.). For example, I clustered and visualized some protein interaction data from UniProt (UniProt Consortium 2011) using Cytoscape (Figure 5.10). An example of a cluster that I identified is shown (Figure 5.10). This cluster contains seven highly connected proteins—they share many more interactions than a typical pair of proteins. A quick gene set enrichment analysis (see Chapter 2) shows that six of these proteins are members of the HOPS complex, which is involved in vacuolar vesicle fusion. (In this case, six of seven is highly significant even after Bonferroni correction, see Chapter 3.) Thus, this known protein complex could have been discovered automatically from the pair wise interaction data. As you can imagine, given that it's possible to turn any high-dimensional data into pairwise interaction data, graph-based clustering is a popular and intuitive clustering approach that scales well to very large data.

5.10 CLUSTERING AS DIMENSIONALITY REDUCTION

Confronted with very high-dimensional data like gene expression measurements or whole genome genotypes, one often wonders if the data can somehow be simplified or projected into a simpler space. In machine learning, this problem is referred to as dimensionality reduction. Clustering can be thought of a simple form of dimensionality reduction. For example, in the case of K-means, each observation can be an arbitrarily high-dimensional vector. The clustering procedure reduces it to being a member of one of the K clusters. If arbitrarily, high-dimensional data can be replaced by simply recording which cluster it was a part of, this is a great reduction in complexity of the data. Of course, the issue is whether the K clusters really capture *all* of the interesting information in the data. In general, one would be very happy if they captured most of it.

EXERCISES

1. The human genome has about 20,000 genes. Initial estimates of the number of genes in the human genome (before it was actually sequenced) were as high as 100,000. How many more pairwise distances would have been needed to do hierarchical clustering of human gene expression data if the initial estimates had turned out to be right?

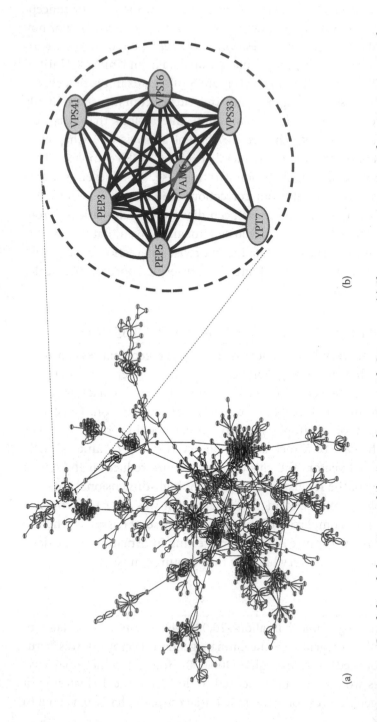

FIGURE 5.10 Graph-based clustering and data visualization with Cytoscape. (a) Shows a representation of protein interaction data as a network or "graph." Each node (gray ovals) in the network represents a protein, and each edge (thin lines) represents a reported interaction (so that multiple connections are possible between any pair of nodes). Clustering of this network using a graph-based clustering method (MCODE, Bader and Hogue 2003) identified 40 clusters. Dashed circle indicates one of the clusters identified. (b) Zoom-in view of a cluster identified shows the large number of reported interactions between these genes.

2. How many distance calculations (as a function of the number of datapoints and clusters) are needed in each iteration to perform the K-means clustering procedure?

3. Online algorithms are iterative procedures that are performed one observation (or datapoint) at a time. These can be very useful if the number of observations is so large it can't be stored in the computer's memory (yes this does happen!). Give an example of an online K-means algorithm. (*Hint*: Consider the "learning signal" provided by a single observation.)

4. Assuming two proteins are 50% identical, what is the expected dependence of the protein sequence distance based on the BLOSUM matrix on sequence length?

REFERENCES AND FURTHER READING

Bader GD, Hogue CW. (2003). An automated method for finding molecular complexes in large protein interaction networks. *BMC Bioinform.* 4(1):1.

Eisen MB, Spellman PT, Brown PO, Botstein D. (December 8, 1998). Cluster analysis and display of genome-wide expression patterns. *Proc. Natl. Acad. Sci. USA* 95(25):14863–14868.

Enright AJ, Van Dongen S, Ouzounis CA. (2002). An efficient algorithm for large-scale detection of protein families. *Nucleic Acids Res.* 30(7):1575–1584.

Henikoff S, Henikoff JG. (1992). Amino acid substitution matrices from protein blocks. *PNAS* 89(22):10915–10919.

de Hoon MJL, Imoto S, Nolan J, Miyano S. (2004). Open source clustering software. *Bioinformatics* 20(9):1453–1454.

Li L, Stoeckert CJ Jr., Roos DS. (2003). OrthoMCL: Identification of ortholog groups for eukaryotic genomes. *Genome Res.* 13:2178–2189.

Mézard M. (February 16, 2007). Computer science. Where are the exemplars? *Science.* 315(5814):949–951.

Rousseeuw PJ. (1987). Silhouettes: A graphical aid to the interpretation and validation of cluster analysis. *Comput. Appl. Math.* 20:53–65.

Shannon P, Markiel A, Ozier O, Baliga NS, Wang JT, Ramage D, Amin N, Schwikowski B, Ideker T. (2003). Cytoscape: A software environment for integrated models of biomolecular interaction networks. *Genome Res.* 13(11):2498–2504.

Strasser BJ. (Winter 2010). Collecting, comparing, and computing sequences: The making of Margaret O. Dayhoff's atlas of protein sequence and structure, 1954–1965. *J. Hist. Biol.* 43(4):623–660.

UniProt Consortium. (2011). Reorganizing the protein space at the Universal Protein Resource (UniProt). *Nucleic Acids Res.* gkr981.

CHAPTER 6

Mixture Models and Hidden Variables for Clustering and Beyond

W E'RE NOW READY TO encounter one of the most elegant and most widely used models in machine learning, the mixture model. In doing so, we will introduce the powerful concept of hidden variables: observations drawn from a pool just like our data, but whose values we didn't actually observe.

It's probably worth noting before we embark on mixtures of Gaussians that the discussion in the context of clustering is largely for pedagogical purposes—although mixtures of Gaussians are widely used for clustering in a variety of types of data analysis in many fields, high-dimensional molecular biology data analysis is not one of them. So if you're looking for a practical approach to clustering your data, this might not be the place to find it. But mixtures of Gaussians are like the lac operon of machine learning. They are one of the places where key concepts are first demonstrated before people move on to bigger and more practical problems. In this chapter, I will use mixtures of Gaussians to introduce the key statistical concept of hidden variables and the expectation–maximization (E-M) algorithm, a very important algorithm for maximizing the likelihood.

In the last sections of this chapter, I'll discuss MEME and Cufflinks (two popular bioinformatics softwares) that have at their hearts mixture models. We will see that mixture models are not limited to Gaussian distributions—they are a general way to model data with complicated statistical properties.

6.1 THE GAUSSIAN MIXTURE MODEL

K-means (which we considered in Chapter 5) is a very widely used clustering method because of its conceptual and algorithmic simplicity. However, it can't be considered a statistical modeling technique because the models are not probability distributions. However, it's quite straightforward to generalize K-means to a probabilistic setting. In doing so, we will finally have our first probabilistic machine learning method. To do so, we simply interpret the K cluster means of K-means as the means of multivariate Gaussian distributions (described in Chapter 4). Formally, this implies that each observation was now drawn i.i.d. from one of several Gaussian component distributions. The interesting part is that we don't know which datapoints were drawn from which of these Gaussians, nor do we know the mean and variance parameters of these Gaussians. Amazingly, if we define the update rule (or algorithm) cleverly, even if we start with a random guess, the computer will be able to automatically learn (or infer) all of this from the data.

One major difference between the mixture of Gaussians and the K-means algorithm is that the datapoints will not be assigned absolutely to one cluster or another. Although in the probabilistic model we assume each datapoint is drawn from one of the component Gaussians that represent each cluster, we consider the "true" assignment to the cluster to be unobserved. Instead of making a so-called "hard assignment" to a cluster, we will assign each datapoint fractionally according to its probability of being part of each cluster. This type of "fuzzy" assignment allows datapoints that are really in between two clusters to be partly assigned to both, reflecting the algorithm's uncertainty about which cluster it really belongs in.

As promised, the mixture of Gaussians is a probabilistic model, and its objective function is a likelihood function. Let's start by writing it down:

$$L = P(X|\theta) = \prod_{i=1}^{n} \sum_{c=1}^{K} P(Z_{ic}=1|\theta)P(X_i|Z_{ic}=1,\theta) = \prod_{i=1}^{n} \sum_{c=1}^{K} \pi_c N(X_i|\mu_c,\Sigma_c)$$

In this formula, we start on the left at the general formula for the likelihood, define the K-component mixture model in general, and finally specify the Gaussian mixture in the last formula on the right. In a Gaussian mixture, each (possibly high-dimensional) datapoint X_i, is drawn from one of K Gaussian distributions, $N(X|\mu, \Sigma)$. I will use an indicator variable, Z, to represent the assignment to each component of the mixture: $Z_{ic} = 1$, if the ith observation is in the cth cluster and 0, if it is not. Notice that I have introduced a so-called "prior" probability $P(Z_{ic} = 1|\theta) = \pi_c$ of drawing an observation from each class from the mixture. The mixture model says, first pick a cluster (or class) according to the prior probability, and then draw an observation X from a Gaussian distribution *given* that class. Because in the mixture model, the class for each datapoint is unknown, we sum (or "marginalize") over the possible classes for each datapoint (hence the sum over c). In the standard formulation of the mixture of Gaussians, this prior is simply a multivariate Bernoulli distribution (also known as a discrete distribution or sometimes a multinomial) whose parameters are the "mixing parameters," π (not to be confused with 3.1415... that occurs in the formula for the Gaussian probability). In order to make the prior a proper probability distribution, these mixing parameters must sum to 1, such that $\sum_{c=1}^{K} P(Z_{ic} = 1|\theta) = \sum_{c=1}^{K} \pi_c = 1$. You can think of these mixing parameters as our expectation about the fraction of datapoints from each cluster.

In the context of probabilistic models, we will interpret the indicator variable as a random variable (just like our observations) but whose values we have not measured: We imagine that some truth exists about whether the ith observation from the pool belongs to the cth cluster, but we simply have not observed it (Figure 6.1). This type of variable is referred to as a "hidden" variable. Note that we have now increased the dimensionality of our data: instead of each datapoint simply being some observations X_i, each datapoint can now be thought of as having an observed component X_i and a "hidden" component Z_i that specifies which cluster it was drawn from.

The interpretation of the cluster assignment as a hidden variable allows us to motivate the specific form that we chose for the mixture of model using the rules of probability. The mixture model can be thought of as writing the probability of the data X as

$$P(X) = \sum_{Z} P(X, Z) = \sum_{Z} P(Z)P(X|Z)$$

where I have left out the parameter vectors and the product over observations for clarity. In some sense, by adding the hidden variables we are making our problem more complicated than it really is: we have a dataset X, and rather than just analyzing it, we are artificially supplementing it

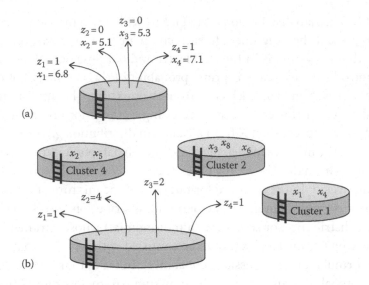

(a)

(b)

FIGURE 6.1 (a) The mixture model adds a hidden indicator variable Z, to each of the observed variables, X. (b) A mixture model can be thought of as drawing observations from a series of pools. First, each an indicator variable, Z, is drawn from a pool that tells us which cluster (or class) each datapoint, X, belongs to. Depending on Z, X is assumed to have different parameters (i.e., depending on which cluster X belongs to). In a clustering situation, the Z is not observed—we have to "fill in" or infer their values.

with hidden variables and using mathematical tricks to keep it all ok. But in fact we are doing this for good reason: Our data X, has a very compli- cated distribution that we cannot model using a standard probability dis- tribution. However, we believe that if we divide up our data into "clusters" (as determined by the hidden variable, Z) each cluster will have a much simpler distribution that we can model (in our case using Gaussians). Said another way, $P(X)$ was too complicated to model directly, but the "class- conditional density" $P(X|Z)$ has a simple form. Therefore, we have traded a difficult distribution of our data for a joint distribution of our data and some hidden variables whose conditional distribution we can character- ize. Figure 6.2a illustrates the idea of trading of a single complicated distri- bution for two Gaussians in one dimension. Figure 6.2b shows an example of how a two-component mixture model can be used to fit single-cell RNA-seq data that we saw in Chapter 2 does not fit very well to standard probability distributions. Mixture models of this type are used in practice to analyze single-cell sequencing data (e.g., Kharchenko et al. 2014). As you can probably imagine, if the number of components in the mixture

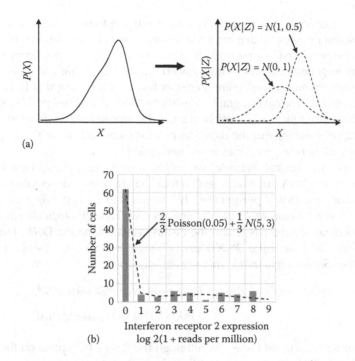

(a)

(b)

FIGURE 6.2 Trading a complicated distribution for a mixture of simple conditional distributions. In (a), the graph on the left shows an asymmetric probability distribution that doesn't fit well to any standard model. The graph on the right shows that the distribution on the left is actually an equal mixture of two very simple Gaussian distributions. In (b), the distribution of single-cell RNA-seq data (discussed in Chapter 2) is shown as gray bars. The dotted trace shows a two-component mixture model (mixture of two-thirds Poisson and one-third Gaussian) that does a reasonably good job of fitting the data. The parameters for each distribution are shown in the parenthesis.

gets large enough, it's possible to represent arbitrarily complicated distributions using the mixture models. In fact, in the computer vision world, Gaussian mixture models with large numbers of components can be used to model everything from hand-written digits on bank cheques to the shapes of bacilli in micrographs of sputum.

BAYESIAN NETWORKS AND GRAPHICAL MODELS

Now that I have introduced the joint distribution of our hidden and observed variables, we can understand what is meant in the machine learning world by "Bayesian networks." In complicated data analysis projects, we might have several variables whose relationships we are trying to understand. The idea of Bayesian networks is that the dependencies among variables can be very usefully represented in a network. Bayesian networks are a specific example of a more general class of models called "graphical" models. A "graph" is what mathematicians and computer scientists call the thing that biologists call a network. Formally, a graph consists of "nodes" and "edges." Bayesian networks are probabilistic models that can be represented as directed, acyclic graphs, meaning that the direction of all the connections in the graph is known and there are no cycles in the network.

To use a biological example, we might want to understand how DNA sequence and RNA expression level affect the chances of developing a disease. From a statistical perspective, this amounts to understanding the joint distribution $P(Disease, DNA, RNA)$. However, we know (biologically) that the RNA depends on the DNA, and the disease depends on the DNA, but not the other way around (the DNA doesn't change depending on the disease or the RNA). So we might try to write something like

$$P(Disease, DNA, RNA) = P(Disease|DNA)P(Disease|RNA)$$

$$\times P(RNA|DNA, Disease)P(DNA)$$

Furthermore, we might know that although the RNA can depend on the disease or the DNA, the effect of each is independent. So we can write

$$P(RNA \mid DNA, Disease) = P(RNA|DNA)P(RNA|Disease)$$

I hope it's clear from this example that, even with only three variables, writing down what we know about the joint distribution can get complicated. Imagine what happens when we start adding hidden variables to this model. However, this complicated structure of dependencies can be represented in a straightforward way using the network (or "graph") in Figure 6.3a. In this

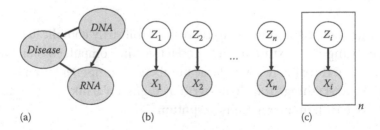

(a)   (b)   (c)

FIGURE 6.3 Graphical model diagrams. These diagrams are used to describe the dependence (and conditional independence) among the variables in a probabilistic model. Arrows connect dependent variables, and gray circles represent observed variables, while unfilled circles represent unobserved (or hidden or latent) variables. See text for details.

representation, shaded circles represent observations of random variables (data). Arrows between circles (or "directed edges") represent connections for which the direction of causality is known: DNA causes differences in RNA expression, and not vice versa. Lines between circles (or "undirected edges") represent connections that could go both ways: RNA levels could cause the disease, but the disease could also cause changes in the RNA levels. Notice that this graphical model does not qualify as a Bayesian network because not all of the edges are directed. It turns out that in general models where all the edges are directed and where there are no cycles in the graph are much more computationally tractable. If possible, people will try to formulate their model as a directed acyclic graph or DAG for short, even if this means making some simplifying assumptions. For example, in order to convert this model into a DAG, we might make the (not entirely true) assumption that RNA levels can affect the disease, but that the disease doesn't affect RNA levels (see Exercises).

We have already seen a simple probabilistic model that does qualify as a Bayesian network: the mixture of Gaussians. We have (multivariate) observations, $X_1, X_2,..., X_n$ and associated hidden variables $Z_1, Z_2,..., Z_n$, specifying which of the K possible Gaussian components (or clusters) that those observations were drawn from. As we defined it, the observations depend on the cluster, but the cluster does not depend on the observations. This means that the arrow points from the Z variables to the X variables as illustrated in Figure 6.3b. I've also adopted the convention that the hidden variables are represented as unfilled circles in the diagram. Finally, because there are typically many observations that are treated as i.i.d. for the Bayesian network, rather than writing all of them out, sometimes people will just put a box around the structure and write the number of times it repeats beside the box as indicated in Figure 6.3c.

6.2 E-M UPDATES FOR THE MIXTURE OF GAUSSIANS

As with the K-means clustering, it should be clear that the objective function for the Gaussian mixture model is very high-dimensional (has many parameters) and therefore will in general be difficult to optimize. For example, in the bivariate case (typically used for illustration purposes, see Figure 6.4 for an example) even with only two clusters there are already 11 parameters, so brute-force exploration of the likelihood surface is already impractical. However, as with K-means, it is possible to derive an iterative procedure (algorithm) that is guaranteed to increase the objective function at each step. In this case, because the objective function is the likelihood, if the algorithm reaches the maximum of the objective function, we will have maximum likelihood estimates (ML estimates) of the parameters. It's an example of the E-M algorithm, a very general learning

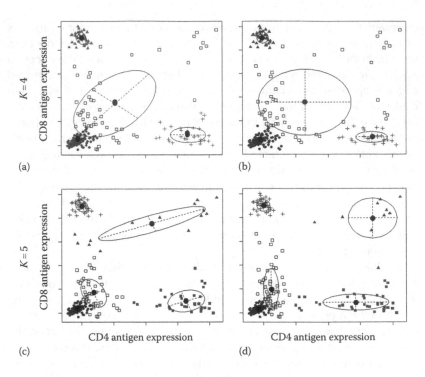

(a) (b) (c) (d)

CD4 antigen expression

CD4 antigen expression

FIGURE 6.4 Clustering using Gaussian mixture models (Fraley et al. 2012) on the CD4 and CD8 antigen expression levels from ImmGen. Black circles identify the means identified by the mixture of Gaussians and ellipses illustrate the covariance matrices. Each gray symbol indicates one cell type, and the shape of the symbols indicates the cluster to which it is most likely to belong to. Notice that because the covariances of the clusters are different, cell types are not necessarily most likely to be assigned to the closest mean (as with k-means). (a, b) show the convergence of E-M assuming 4 clusters, and (c, d) show the convergence with 5 clusters. (a, c) Show a Gaussian mixture model with full covariance, while (b, d) show the Gaussian Mixture model with diagonal covariance. As with the K-means example (in Chapter 5) the Gaussian Mixture model has a difficult time identifying the small CD4+CD8+ cluster with $K = 4$, preferring instead to break up the large, dispersed CD4−CD8− into two clusters. With $K = 5$ it does seem to correctly identify the small cluster.

(or inference) algorithm for probabilistic models with hidden variables. The general procedure outlined for deriving the E-M algorithm can be applied to a large class of probabilistic models with hidden variables.

Because the derivation of the E-M algorithm is a bit long, I will start by giving the answers. We start by "assigning" each datapoint to a cluster.

Since the variable Z represents the assignment of each datapoint to a cluster, our best guess to which cluster the datapoint belongs is the expectation (or average) of the hidden variable, Z. This is the so-called "E-step" or "expectation-step" of the E-M algorithm. Because we have interpreted the indicator variable as a random (albeit hidden) variable, we assign a probability distribution to this variable and compute its expectation, which we will indicate as $\langle Z \rangle$. The angle brackets are a notation for averages or expectations widely used in physics.

$$\langle Z_{ik} \rangle \equiv E[Z_{ik}] = \frac{P(X_i | Z_{ic} = 1, \theta) P(Z_{ic} = 1 | \theta)}{P(X_i | \theta)} = \frac{\pi_k N(X_i | \mu_k, \Sigma_k)}{\sum_{c=1}^{K} \pi_c N(X_i | \mu_c, \Sigma_c)}$$

Again, the first formula is for the general case of the mixture model, and the second is for the mixture of Gaussians. Another way of looking at the E-step is that we "fill in" or infer the values of the hidden variables by assigning them to their expected values. Because in this case, the hidden variables are 0 or 1 indicator variables, computing their expectations corresponds to inferring the (posterior) probability that the observation i belongs to the cluster k (the expectation of a 0 or 1 variable is just the probability of observing 1).

After assigning datapoints to clusters by filling in the hidden variables, we re-estimate the parameters to maximize the objective function. This is the so-called maximization or "M-step." For the mixture of Gaussians, we have the following formulas:

$$\mu_k = \frac{1}{\sum_{i=1}^{n} \langle Z_{ik} \rangle} \sum_{i=1}^{n} \langle Z_{ik} \rangle X_k$$

This is the "update equation" for the mean of the Gaussians. Notice the similarity of this equation to the formula for the means of K-means. We have a weighted average, where the weights are the expected assignments of each datapoint to each cluster. We also need a formula to calculate the mixing parameters, or the relative fractions of each Gaussian.

$$\pi_k = \frac{1}{n} \sum_{i=1}^{n} \langle Z_{ik} \rangle$$

The interpretation of this formula is simply that it's the expected number of datapoints assigned to the cluster k, divided by the total number of datapoints. This is just the "expected" fraction of the datapoints that are assigned to each cluster. Finally, we need a formula for the covariance matrices:

$$\Sigma_k = \frac{1}{\sum_{i=1}^{n} \langle Z_{ik} \rangle} \sum_{i=1}^{n} \langle Z_{ik} \rangle (X_i - \mu)(X_i - \mu)^T$$

which again can be interpreted as a weighted version of the standard formula for the maximum likelihood estimations (MLEs) of the covariance. Once we've re-estimated the parameters at the M-step, we just go back to the E-step and fill in the hidden variables again.

Thus, the implementation of E-M for the mixture of Gaussians is just a matter of calculating these four formulas, which in many cases is quite straightforward using a package like MATLAB® or R. The algorithm alternates between filling in the expected assignments for each datapoint and then optimizing the parameters of the clusters and the mixing parameters. Eventually, the assignments and parameters stop changing and the algorithm has "converged" to a local maximum in the likelihood.

Although the E-M algorithm is guaranteed to converge to a local maximum in the likelihood, there is no guarantee that a mixture model will have only one local optimum. Typically, several initializations of the parameters are chosen and the algorithm is run to convergence. Sometimes, the initialization will be done based on another clustering algorithm, like K-means to make sure that the initial guesses are not too far off. After a few different initializations are tried, the likelihoods under the final parameters for each initialization are compared, and the best parameters are chosen as the final answer.

In general, the E-M algorithm for any mixture of distributions is relatively simple to implement. The formulas for the hidden variable and mixing parameter are the same as mentioned earlier, and the updates for the parameters are typically weighted versions of the standard MLEs for the parameters. This is one of the reasons that it is a favorite optimization algorithm for mixture models. Note that this does not imply that E-M is the "best" or "fastest" way to optimize the likelihood—in most cases it's neither. But in many cases, it is the simplest optimization strategy to implement, so if you have a probabilistic model that you want to fit, and you can derive E-M updates, this is probably where you want to start.

6.3 DERIVING THE E-M ALGORITHM FOR THE MIXTURE OF GAUSSIANS

The E-M algorithm is a very general strategy to derive iterative optimization procedures to maximize the likelihood. The key idea of the E-M algorithm is that we will not even try to find analytical formulas for the parameters by solving $\partial/\partial\theta \log L = 0$. Instead, we will replace the likelihood function with an easier function to optimize, where some of the unknown variables have been fixed. We will then alternate between maximizing the function (M-step) and filling in the unknown variables with our best guess or "expected" values (E-step). The amazing result (which is beyond the scope of this book) is that if the objective function that we maximize is chosen cleverly, we will be guaranteed to also maximize the likelihood at each iteration.

The alternative function that we will optimize for the purposes of this book is the so-called "expected complete log-likelihood" following Michael Jordan's interpretation of E-M (Jordan and Jacobs 1994). To write down this function we first define the "complete likelihood," namely, the likelihood of the data *if we knew which cluster each datapoint belonged to*. This complete likelihood, CL, for the mixture model is

$$CL = \prod_{i=1}^{n}\prod_{c=1}^{K}\left[P(Z_{ic}=1|\theta)P(X_i|Z_{ic}=1,\theta)\right]^{Z_{ic}} = \prod_{i=1}^{n}\prod_{c=1}^{K}\left[\pi_c N(X_i|\mu_c,\Sigma_c)\right]^{Z_{ic}}$$

where once again the first formula is the general case for all mixture models, and the second formula is for the mixture of Gaussians. To write this formula, I used the indicator variable Z that has the value 1 if the ith datapoint belonged to the cth cluster and 0 otherwise. Notice the effect this indicator variable has: All the terms in the product that correspond to clusters to which the ith datapoint does not belong will be raised to the power of 0 (they will be set to 1). The terms that correspond to the cluster to which the ith datapoint belongs will be raised to the power of 1 and will be included in the product. Again, the difference between this "complete" likelihood and the actual likelihood is that to write this formula, we need to know the "Z" indicator variables that tell us which cluster every datapoint belongs to.

Although this new objective function might not look much simpler, we can now proceed to take its logarithm and expectation over the hidden variables

$$\log CL = \sum_{i=1}^{n} \sum_{c=1}^{K} Z_{ic} \left[\log \pi_c + \log N(X_i | \mu_c, \Sigma_c) \right]$$

$$\langle \log CL \rangle = \sum_{i=1}^{n} \sum_{c=1}^{K} \langle Z_{ic} \rangle \left[\log \pi_c + \log N(X_i | \mu_c, \Sigma_c) \right]$$

where in the second step we use the linearity of the expectation operator to move it inside the sum. Given what you know from the previous sections about derivatives and logarithms of Gaussian distributions, you probably can already guess that it's now possible to solve $\partial / \partial \theta \langle \log CL \rangle = 0$ and optimize the expected complete log-likelihood. For example, we will start with the mean parameters

$$\frac{\partial \langle \log CL \rangle}{\partial \mu_k} = \sum_{i=1}^{n} \langle Z_{ik} \rangle \frac{\partial \log N(X_i | \mu_k, \Sigma_k)}{\partial \mu_k}$$

$$= \sum_{i=1}^{n} \langle Z_{ik} \rangle \frac{\partial}{\partial \mu_k} \left[-\frac{1}{2} (X_i - \mu_k)^T \Sigma_k^{-1} (X_i - \mu_k) \right] = 0$$

First, notice that I didn't write the sum of all the clusters in this formula: the derivative with respect to the mean of the kth cluster implies that all the terms that don't depend on μ_k will have zero derivatives and therefore not matter for the optimization. So the terms related to all the other clusters won't matter for the kth mean. Similarly, in the second step, I haven't bothered to write the rest of the terms of the (log) Gaussian distribution because they don't depend on μ_k and therefore will disappear. To simplify the derivation, we will use the matrix calculus trick that $(\partial / \partial x)[x^T A x] = x^T (A + A^T)$, where A is a matrix, and the derivative is with respect to a whole vector, x. Because the covariance matrix is symmetric, it equals its transpose, so we have

$$\frac{\partial \langle \log CL \rangle}{\partial \mu_k} = \sum_{i=1}^{n} \langle Z_{ik} \rangle \left[-\frac{1}{2} (X_i - \mu_k)^T 2 \Sigma_k^{-1} \right] = \sum_{i=1}^{n} \langle Z_{ik} \rangle (X_i - \mu_k)^T \Sigma_k^{-1} = 0$$

As before, we can multiply each term of the sum and the 0 vector by the covariance matrix, which gives.

$$\frac{\partial \langle \log CL \rangle}{\partial \mu_k} = \sum_{i=1}^{n} \langle Z_{ik} \rangle (X_i - \mu_k)^T = 0$$

This can be solved to get the update formula for the mean I gave above. To derive the formula for the mixing parameters, we have to take into account the constraint that they have to sum to 1. We add this to the objective function using a Lagrange multiplier. The constraint can be written as $1 - \sum_{c=1}^{K} \pi_c = 0$, so the expected complete log-likelihood with the constraint is

$$\langle \log CL \rangle = \sum_{i=1}^{n} \sum_{c=1}^{K} \langle Z_{ic} \rangle \left[\log \pi_c + \log N(X_i | \mu_c, \Sigma_c) \right] + \lambda \left(1 - \sum_{c=1}^{K} \pi_c \right)$$

Since the Gaussian portion of the objective function doesn't depend on the mixing parameter, its derivatives will be zero, as will the terms in the sums that don't depend on the kth mixing parameter. We get

$$\frac{\partial \langle \log CL \rangle}{\partial \pi_k} = \sum_{i=1}^{n} \frac{\langle Z_{ik} \rangle}{\pi_k} - \lambda = 0$$

which we can solve to give

$$\pi_k = \frac{1}{\lambda} \sum_{i=1}^{n} \langle Z_{ik} \rangle$$

Remember that we will get a similar formula for each of the K-mixing parameters. To figure out what λ is, we substitute the update equation into the constraint

$$1 = \sum_{c=1}^{K} \pi_c = \sum_{c=1}^{K} \frac{1}{\lambda} \sum_{i=1}^{n} \langle Z_{ic} \rangle = \frac{1}{\lambda} \sum_{i=1}^{n} \sum_{c=1}^{K} \langle Z_{ic} \rangle = \frac{1}{\lambda} \sum_{i=1}^{n} 1 = \frac{1}{\lambda} n$$

which means that $\lambda = n$. Notice that in going from the third to the fourth formula (starting from the left) I used the fact that the sum of the $\langle Z \rangle$ over the clusters must also sum to 1. This is because each datapoint must be fully assigned; it's just that we don't know which cluster it belongs to. Another way to look at it is that we imagine that the datapoint must have belonged to *one* of the Gaussians.

Deriving the updates for the covariance involves the same sort of matrix calculus tricks we saw in Chapter 4. Overall, although the derivation is very similar to what we have seen for MLE, remember, the "M" or maximization step does not actually maximize the likelihood function. Instead, we are maximizing a simpler objective function, the expected complete log-likelihood. The magic is that by repeatedly maximizing the expected complete log-likelihood and filling in the hidden variables with their expectations (the "E" step) we are always increasing the likelihood. By running this process until convergence, we are guaranteed to find a maximum in the likelihood.

Another important comment here is that for the mixture of Gaussians we were not able to derive closed form solutions to maximize the likelihood, but we *were* able to derive analytic solutions for the M-step. For more complicated models, even this might not be possible. We might only find a formula that *increases* (not maximizes) the expected complete log-likelihood at each iteration, or instead we might give up on formulas and only numerically maximize (or increase) the expected complete log-likelihood. The E-M algorithm also permits us to include analytic updates for some of the parameters, and mix these with numerical updates for others where we can't solve the equations. Sometimes, these kinds of algorithms will be called "generalized" E-M or sometimes just E-M (even though they might not actually be doing "M" at the M-step). Amazingly, they will all work: we will be guaranteed to be increasing the likelihood at each step.

To find the update for the E-step, we use Bayes' theorem to fill in the expectation of the hidden variables. We have

$$\langle Z_{ik} \rangle = P(Z_{ik} = 1 | X, \theta) = \frac{P(X | Z_{ik} = 1, \theta) P(k | \theta)}{P(X | \theta)}$$

The first equality uses the fact that the expectation for a Bernoulli variable is just the probability of the positive outcome, and the second is Bayes' theorem. To simplify this formula, we note that for all the terms relating to datapoints other than i, $P(X | Z_{ik} = 1, \theta) = P(X | \theta)$, so those terms will cancel out:

$$\langle Z_{ik} \rangle = \frac{\pi_k N(X_i | \mu_k, \Sigma_k) \prod_{j \neq i}^{n} \sum_{c=1}^{K} \pi_c N(X_j | \mu_c, \Sigma_c)}{\sum_{c=1}^{K} \pi_c N(X_i | \mu_c, \Sigma_c) \prod_{j \neq i}^{n} \sum_{c=1}^{K} \pi_c N(X_j | \mu_c, \Sigma_c)}$$

$$= \frac{\pi_k N(X_i | \mu_k, \Sigma_k)}{\sum_{c=1}^{K} \pi_c N(X_i | \mu_c, \Sigma_c)}$$

where the last step is the formula given above for the E-step.

6.4 GAUSSIAN MIXTURES IN PRACTICE AND THE CURSE OF DIMENSIONALITY

Despite their conceptual and mathematical elegance, the general Gaussian mixture models are almost never used in practice for clustering high-dimensional data. The reason for this is simple: there are simply too many parameters in these models to fit them. The reason that the number of parameters gets so large is an example of the so-called "curse of dimensionality." Let's calculate the number of parameters in the likelihood function of the Gaussian mixture model with K components and d dimensions. For each component, we have a mean vector with d-parameters, and a symmetric covariance matrix with $(d + 1)d/2$ parameters. We also need $K - 1$ mixing parameters. In all, we're looking at an objective function with $K - 1 + Kd(1 + (d/2 + 1/2))$ or more than $(1/2)Kd^2$ parameters.

For 10 clusters of 100-dimensional data, we're already looking at more than 50,000 parameters. It's just too hard to optimize: Although the E-M algorithm is guaranteed to give us a local optimum in the likelihood, there's just no way to try enough random starting values to convince ourselves that we've actually found the maximum likelihood. Furthermore, even if we give up on E-M and start implementing some fancier optimization methods, we need thousands of 100-dimensional observations to have enough data to estimate all these parameters.

In practice, we reduce the complexity of the likelihood function by placing constraints on the covariance matrices. This reduces the number of parameters (see Exercises). In the limit where covariance is the identity matrix (I), we have something very much like K-means. A very interesting case is where $\Sigma = sI$, where s is a vector of variances. This allows each cluster to have different spreads, and also to weigh the dimensions differently. In practice, this can be very powerful if the observations are not equally spread out in the high-dimensional space. Figure 6.3 illustrates the effect of the diagonal covariance assumption on the model (compare a and c to b and d). At least for this two-dimensional example, the assumption of diagonal covariance does not greatly impact which clusters are found.

6.5 CHOOSING THE NUMBER OF CLUSTERS USING THE AIC

To choose the "best" clustering and the optimal number of clusters for the Gaussian Mixture, we could go ahead and use a measure like the silhouette that we introduced in the previous chapter. However, one attractive feature of the Gaussian mixture model is that the objective function is the likelihood. At first, we might guess that we could choose the number of clusters that maximizes the likelihood. However, we know that the likelihood will always increase as the number of clusters increases, because adding additional clusters increases the number of parameters (see the previous section). So we can't choose the number of clusters by maximum likelihood. Fortunately, under certain assumptions, it's possible to predict how much the fit of a model will improve if the extra parameters were only fitting to the noise. This means we can trade off the number of parameters in the model (the "complexity of the model") against the likelihood (or "fit of the model"). For example, for a model with k parameters and maximum likelihood L, the AIC (Akaike Information Criterion) is defined as

$$AIC = 2k - 2\log(L)$$

Information-theoretic considerations suggest that we choose the model that has the minimum AIC. So if a more complex model has one additional parameter, it should have a log-likelihood ratio of at least one (compared to the simpler model) before we accept it over a simpler model. Thus, the AIC says that each parameter translates to about 1 units of log-likelihood ratio. The AIC assumes the sample size is very large, so in practice, slightly more complicated formulas with finite sample corrections are often applied. To use these, you simply try all the models and calculate the (finite sample corrected) AIC for all of them. You then choose the one that has the smallest (Figure 6.5).

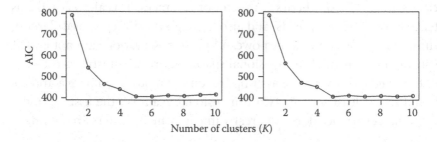

FIGURE 6.5 Choosing models (cluster number) using the AIC. For the CD4–CD8 data shown here, the AIC reaches a minimum with 5 clusters for both the Gaussian mixture model with full covariance (left) with diagonal covariance (right). This agrees with the observation that the CD4+CD8+ cells were not captured by the models with only four clusters.

6.6 APPLICATIONS OF MIXTURE MODELS IN BIOINFORMATICS

Although mixtures of Gaussians aren't usually used in practice to cluster genome-scale expression datasets, mixture models have proven very effective at modeling another type of important molecular biology data: sequences. Because sequence data is not continuous, usually Gaussian models are not appropriate. We'll consider two popular bioinformatics applications that are based on mixture models, however, in both cases they will turn out to be mixtures of other distributions.

The MEME software (Bailey and Elkan 1995), for identifying "motifs" or patterns in DNA and protein sequences, is based on a two-component mixture model. In this case, the two "clusters" are (1) a family of short sequences that share a similar pattern, and (2) all the rest of the "background" sequences. Amazingly, MEME can effectively separate out the short sequences that share a pattern without needing to know what the pattern is in advance: The pattern is the parameters of the cluster that are estimated by E-M.

MEME starts by choosing a "motif width" w, which will be the length of the pattern it will discover. The input DNA or protein sequence (say of length n) is the decomposed into subsequences of length w, "w-mers." These w-mers can then be treated as i.i.d. observations drawn from a pool where some of the w-mers are members of the motif cluster, and the others are members of the background cluster. As in any clustering problem, we don't know which of the w-mers belong in which cluster ahead of time. Once again, we can write down the likelihood of the data:

$$L = P(X|\theta) = \prod_{i=1}^{n-w+1} \left[\varphi P(X_i|f) + (1-\varphi)P(X_i|g) \right]$$

In writing the objective function for this mixture model, I didn't include a sum over the mixture components, because in this case there are only two: either a w-mer is in the motif cluster, or in the background class. Instead of writing the sum, I explicitly wrote out the two components, and gave the motif cluster the parameters f and mixing parameter φ, and the background cluster parameters g and mixing parameter $1-\varphi$. Once again, we will think of each position, j, in the w-mer as a vector X_j, where the bth component of X_j is 1 if the sequence has the DNA base (or protein residue) b at that position. For example, the w-mer TCCACC would be

$$X_{15} = \begin{matrix} 0 & 0 & 0 & 1 & 0 & 0 \\ 0 & 1 & 1 & 0 & 1 & 1 \\ 0 & 0 & 0 & 0 & 0 & 0 \\ 1 & 0 & 0 & 0 & 0 & 0 \end{matrix}$$

where the bases have been ordered arbitrarily from top to bottom as, A, C, G, T. A protein sequence would be represented as a $20 \times w$ matrix.

To write the probability of a w-mer under the motif or background model, MEME assumes that

$$P(X_i|f) = \prod_{j=1}^{w} \prod_{b \in \{A,C,G,T\}} f_{mjb}^{X_{jb}}$$

where f is a matrix of parameters for the motif cluster specifying the probability of observing base b at position j. The matrix of probabilities, f, defines that pattern of sequences that would be expected if we were to draw w-mers that were all from the motif family. Similarly, the matrix g defines the w-mers that we would expect from the background pool. For example, a parameter matrix for the motif might look like

$$f = \begin{matrix} 0.083 & 0.667 & 0.083 & 0.75 & 0.667 & 0.167 \\ 0.083 & 0.083 & 0.083 & 0.083 & 0.083 & 0.25 \\ 0.75 & 0.167 & 0.083 & 0.083 & 0.167 & 0.417 \\ 0.083 & 0.083 & 0.75 & 0.083 & 0.083 & 0.167 \end{matrix}$$

Thus, MEME is based on a mixture model, but it is not a mixture of Gaussians. Instead, it is a mixture of categorical (or sometimes called multinomial) probability models that were introduced in Chapter 4. Let's take a minute to consider what the formula for the categorical probability model says. For example, it gives

$$P(CACGTG|f) = f_{1A}^0 f_{1C}^1 f_{1G}^0 f_{1T}^0 f_{2A}^1 f_{2C}^0 f_{2G}^0 f_{2T}^0 f_{3A}^0 f_{3C}^1 f_{3G}^0 f_{3T}^0 \cdots$$

Since any number raised to the power of 0 is 1, we get

$$P(CACGTG|f) = f_{1C} f_{2A} f_{3C} f_{4G} f_{5T} f_{6G} = 0.083 \times 0.667 \times 0.083 \times 0.083$$

$$\times 0.083 \times 0.417$$

So the formula says that the probability of a sequence is simply the product of the probabilities of the bases we observed, which I hope you'll agree makes sense.

The clever representation of sequences as matrices of 0s and 1s will also make the objective function look relatively simple for this model. Starting with the complete log-likelihood

$$LC = P(X|\theta) = \prod_{i=1}^{n} \left[\varphi P(X_i|f)\right]^{Q_i} \left[(1-\varphi)P(X_i|g)\right]^{1-Q_i}$$

where I have introduced Q to represent the hidden variable determining which of the two clusters (motif, m or background, g) to which the w-mer, X, belongs (see Figure 6.6). Substituting the discrete probability model mentioned earlier and taking the log and the expectation we get

$$\langle \log LC \rangle = \sum_{i=1}^{n} \langle Q_i \rangle \log \varphi + \langle Q_i \rangle \sum_{j=1}^{w} \sum_{b \in \{A,C,G,T\}} X_{jb} \log f_{jb} + \left(1 - \langle Q_i \rangle\right) \log(1-\varphi)$$

$$+ \left(1 - \langle Q_i \rangle\right) \sum_{j=1}^{w} \sum_{b \in \{A,C,G,T\}} X_{jb} \log g_{jb}$$

We can now go ahead and differentiate to obtain the E-M updates for this model.

In this model, the positions, i, where $Q_i = 1$ are the unobserved "true" positions of the motifs. Although we don't observe these, we fill them in

```
X = GTTGATAACGAGTTTCCACCTTATCACTTATCACTAGTGCTAATCAAACAGCAAAGAATGCTTGATAGAA
Q = 0001000000000000000000000000000000000000000000000000000000000001000000
X₁ GTTGAT
X₂ TTGATA
X₃  TGATAA
    GATAAC
    ATAACG
```

<table>
<tr><td></td><td></td></tr>
<tr><td>X_{n-w-1}</td><td>TGATAG</td></tr>
<tr><td>X_{n-w}</td><td>GATAGA</td></tr>
<tr><td>X_{n-w+1}</td><td>ATAGAA</td></tr>
</table>

FIGURE 6.6 Representation of a DNA sequence, X, using a mixture model. Hidden variables Q indicate whether each position is drawn from the motif (1) or background (0). Here, the DNA sequence is decomposed into w-mers, which we treat as multivariate observations.

with their "expectations" $\langle Q_i \rangle$. If we want to know the likely positions of the motif instances, we should look for the position in the sequence where $\langle Q_i \rangle$ approaches 1. However, the model parameters, f that we estimate are perhaps of even more interest. These represent the sequence preferences of the transcription factor at each position: this is the pattern or "motif."

Here, I have described the simplest version of the MEME model, where all of the data is treated as a simple two-component mixture. However, in practice sequence data might have more structure. For example, we might have 28 bound genomic regions identified in a ChIP-exo experiment: in this case, we might want to find a pattern that occurs once in each sequence fragment (one occurrence per sequence model, oops). In the MEME software, several such model structures have been implemented (Bailey 2002).

A second example of popular bioinformatics software that is based on a mixture model is Cufflinks (Trapnell et al. 2010). RNA-seq experiments aim to detect the mRNA abundance based on the number of short reads mapping to that gene relative to all others in an RNA sequence library. However, in practice, in complex eukaryotic genomes, the reads obtained from each gene might result from several different alternative splice forms. The expression of a gene should be related to the abundance of all of the transcripts that the gene can produce, rather than the average number of reads overlapping all the exons. Usually, however, we aren't sure which splice forms are expressed in a given experiment, and many reads could represent multiple transcripts from the same gene. Cufflinks deals with this uncertainty by treating the transcript as a hidden variable, such that the alignment of reads that we observe is explained by a mixture, where each component (or class) corresponds to one of the known transcripts for that gene. Specifically, we can write a probabilistic model for the alignment of reads as

$$L = P(R|\theta) = \prod_{i=1}^{n} \sum_{t=1}^{K} P(Z_{it} = 1|\theta) P(R_i | Z_{it} = 1, \theta)$$

As usual, I have used Z to represent the hidden variable that tells us which transcript, t, is responsible for the read, R_i. Here, we assume that we know the number of transcripts, K, in advance. Notice that this model corresponds exactly to the mixture model, where the number of reads in alignment i is a mixture of reads from all the transcripts.

Similar to the equation for the mixture of Gaussians, the first part of the likelihood represents the mixing parameter, in this case, the probability that it was transcript t that generated the read R_i. If we assume that reads are generated simply proportional to the amount of RNA present, the probability of observing a read from this transcript is just the fraction of the RNA in this transcript. The amount of RNA in each transcript depends both on the abundance of that transcript, ρ_t, and its length l_t.

$$P(Z_{it} = 1|\rho) = \frac{l_t \rho_t}{\sum_{u=1}^{K} l_u \rho_u}$$

In this model, we assume that we know the lengths of each transcript (based on the genome annotation) but that the abundances are unknown—in fact, these abundances ρ are exactly the parameters that we would like to use the model to estimate.

The second part of the likelihood is the probability of observing a read *given that we know which transcript it came from*. In practice, there can be several attributes of the read that we consider when we compute this probability (Figure 6.7). Most obvious is whether the reads map only to exons that are actually contained in this transcript: the probability of generating a read mapping to exons that are not found in this splice form must be zero. Similarly, if a portion of the read hangs over the edge, it is unlikely that the read was generated from this transcript.

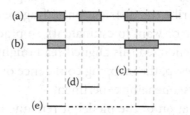

FIGURE 6.7 Factors influencing the probability that a read was part of a transcript. (a and b) Represent the exons (gray bars) of two alternative transcripts at a locus. (c) Shows an aligned read (black line) that could have been generated by either transcript. (d) An aligned read that almost certainly came from transcript (a). (e) A paired-end read that could have come from either transcript, but is more likely to come from transcript (a) because the distance between the paired reads implied by transcript (a) is more consistent with the typical space between paired-end reads (dashed line).

Given that a read falls within a transcript, the simplest model is to assume a uniform probability of observing a read at each possible position in the transcript t:

$$P(R_i|Z_{it}=1) = \frac{1}{l_t - W_i + 1}$$

where W_i represents the length of the read. Thus, in this simplest form, cufflinks assumes that the reads are drawn from a mixture of uniform distributions, with each transcript producing a slightly different uniform distribution.

In practice, cufflinks considers several other more subtle features: for example, for reads where we have sequenced two ends (paired-end reads), we can consider the implied length of the DNA fragment that produced the pair, based on the mapping of the two paired reads in this transcript (Figure 6.7). If this transcript implies a DNA fragment length that is much larger or smaller than typical for our library, we should consider these read alignment less likely given this transcript. In this case, we can write

$$P(R_i|Z_{it}=1,\theta) = \frac{1}{l_t - W_{it} + 1} P(W_{it}|\theta)$$

where I have now also indexed W by the transcript, t, and included some parameters θ to describe the distribution of fragment lengths (we might know this in advance or want to estimate it). In practice, the probability that this transcript generated this aligned read might also depend on the transcript abundances, ρ, the mapping confidence of the reads, and other various biases in the sequencing experiment.

Given the specification of the model, it remains to estimate the transcript abundances, ρ, based on this mixture model. In practice, structure is imposed so that each locus is treated independently and the optimization of the model is performed using numerical methods. Cufflinks is unusual for a bioinformatics software in that it doesn't compute a single estimate of ρ, but rather reports a Bayesian description of the distribution (obtained from a sampling procedure—a type of inference algorithm that we won't cover in this book).

INTEGRATING DATA SOURCES WITH SEQUENCE AND EXPRESSION CLUSTERS

A good illustration of the conceptual power of probabilistic modeling is that models for seemingly disparate data, like gene expression (which are high-dimensional, positive, real numbers) and sequences (which are discrete strings of letters) can be combined. For example, a classic genomics problem is to find genes that have similar expression patterns that also share sequence features (such as transcription factor binding sites) in the noncoding DNA flanking the gene. Using the clustering methods described in Chapter 4 (hierarchical or k-means), you first cluster the expression data, and then in a second step try to find sequence patterns in the noncoding DNA (perhaps using MEME or another motif-finding method). However, with a probabilistic model, it's possible to combine the two types of data into a single model (Holmes and Bruno 2000). We might imagine grouping genes into clusters both based on their expression patterns *and* based on the presence of similar sequence motifs flanking the genes. Let's say the expression data is X and the sequence data is Y, a model for both is

$$L = P(X,Y|\theta) = \prod_{i=1}^{n}\sum_{c=1}^{K}P(c|\theta)P(X_i|c,\theta)P(Y_i|c,\theta) = \prod_{i=1}^{n}\sum_{c=1}^{K}\pi_c N(X_i|\mu_c,\Sigma_c)\prod_{j=1}^{m-w}M(Y_{ij}|f_c,\varphi_c)$$

where
   $N(X)$ is a Gaussian model for the gene expression measurements
   $M(Y)$ is a MEME-like mixture model for the all the m w-mers found in the flanking noncoding DNA sequences, where each cluster is associated with its own sequence model, f_c, and mixing parameter φ_c

In this model, we are saying that the expression and sequence data are dependent, that is, $P(X, Y)$ is not equal to $P(X)P(Y)$, but that *given* that we know what cluster they are truly from, then they are independent. In other words, the expression data and the sequence data depend on each other *only* through the hidden variable that assigns them to a cluster. Slightly more formally, we are saying

$$P(X,Y) = \sum_{Z}P(Z)P(X|Z)P(Y|Z) = \sum_{Z}P(Z)P(X|Z)\sum_{Q}P(Y|Q)P(Q|Z)$$

where I have omitted the products over i.i.d. datapoints to highlight the dependence structure of the model. Notice the effect that this will have when we try to differentiate the expected complete log-likelihood:

$$\langle \log CL \rangle = \sum_{i=1}^{n}\sum_{c=1}^{K}\langle Z_{ic} \rangle \left[\log \pi_c + \log N(X_i|\mu_c,\Sigma_c) + \sum_{j=1}^{m-w}\log M(Y_{ij}|f_c,\varphi_c) \right]$$

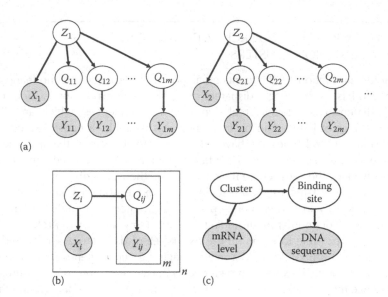

(a)

(b)   (c)

FIGURE 6.8   The Holmes and Bruno joint model for expression data and sequence motifs represented as a graphical model. (a) The structure of the model where all variables are explicitly represented. Z represents the hidden "cluster assignment" and Q represents the hidden "motif" variable. X represents the observed expression levels, and Y represents the observed DNA residues. (b) The collapsed representation where the mDNA residues for each gene are shown in a box and the n genes are shown in another box. (c) The structure of the model with the biological interpretation of each variable.

The terms involving the sequence data Y will be separate from the terms involving the expression data X, so it will be straightforward to derive the E-M updates for this problem as well.

It is instructive to consider the implied causality relationships in this model. To do so, let's use the graphical models representation introduced here. In this representation (Figure 6.8), the hidden variables Z, specify which cluster gene i belongs to. Then, given that cluster, expression data X and some hidden variables Q that represent the presence of motifs in the DNA sequence. These hidden Q variables specify whether a w-mer, Y at position, j, was drawn from the kth motif class or the background class. Remember that both expression data X, and w-mers Y are actually multivariate observations. In the model, we didn't specify what the "cluster" or "Z" represents, but as we look at the structure, it is tempting to interpret the "clusters" as "regulators" that lead to both patterns of gene expression and appearance of binding sites in DNA. I hope this example illustrates the power of the graphical models representation: The structure of the model is much easier to see here than in the formula for the joint distribution.

EXERCISES

1. What is the difference between clustering using a Gaussian mixture model and K-means using the Malhalanobis distance?

2. Derive the update equation for the mixing parameter for the two component mixture of sequence model. (*Hint*: Because I've written the likelihood using one parameter for the two components, you don't have to bother with Lagrange multipliers. Just take logs, expectations, and differentiate directly.)

3. Show that the expectation for a Bernoulli variable (or any 0 or 1 variable) is just the probability of the positive outcome. In other words: $E[X] = P(X = 1)$ if X is $\{0, 1\}$. (*Hint*: Just plug in the values for the Bernoulli distribution into the standard formula for the expectation $E[]$ [given in Chapter 2].)

4. How many parameters are there in a mixture of Gaussians model with diagonal covariances (as a function of K and d)?

5. Show that the discrete probability model $p(X|f)$ used for sequence by MEME is a valid probability distribution, such that the sum over all sequences of length w is 1.

6. Notice that the causality in the Holmes and Bruno model is different from the biological causality: biologically, the binding to DNA would recruit the regulator, which would then cause expression changes. Draw the graphical model representation of this model and write out the formula for the joint distribution including the dependence structure.

7. Draw a graphical models representation of the model where DNA affects RNA and both DNA and RNA affect a disease, but the disease doesn't affect either DNA or RNA. Draw the graphical model representation of this model and write out the formula for the joint distribution including the dependence structure.

8. Hidden variables are widely used in evolutionary modeling to represent unobserved data from ancestral species that are not observed. For example, if we use A to represent the unobserved ancestral data, and X and Y to represent observed data from two extant, descendants of A, the joint probability of the observed data can be written

$P(X,Y) = \sum_A P(X|A)P(Y|A)P(A)$. Draw a graphical models diagram for this probabilistic model. What is $P(A)$ in this formula?

9. Explain the assumption about evolution along each branch being made in the model in question 8.

REFERENCES AND FURTHER READING

Bailey TL. (2002). Discovering novel sequence motifs with MEME. *Curr. Protoc. Bioinform./Editorial Board, Andreas D. Baxevanis... [et al.]* Chapter 2:Unit 2.4.

Bailey TL, Elkan C. (1995). The value of prior knowledge in discovering motifs with MEME. *Proceedings of the International Conference on Intelligent Systems for Molecular Biology (ISMB)*, Cambridge, U.K., Vol. 3, pp. 21–29.

Fraley C, Raftery AE, Murphy TB, and Scrucca L. (2012). mclust Version 4 for R: Normal mixture modeling for model-based clustering, classification, and density estimation. Technical Report No. 597. Department of Statistics, University of Washington, Seattle, WA.

Holmes I, Bruno WJ. (2000). Finding regulatory elements using joint likelihoods for sequence and expression profile data. *Proceedings of the International Conference on Intelligent Systems for Molecular Biology (ISMB)*, La Jolla, CA., Vol. 8, pp. 202–210.

Jordan MI, Jacobs RA. (1994). Hierarchical mixtures of experts and the EM algorithm. *Neural Comput.* 6(2):181–214.

Kharchenko PV, Silberstein L, Scadden DT. (July 2014). Bayesian approach to single-cell differential expression analysis. *Nat. Methods.* 11(7):740–742.

Trapnell C, Williams BA, Pertea G, Mortazavi A, Kwan G, van Baren MJ, Salzberg SL, Wold BJ, Pachter L. (2010). Transcript assembly and quantification by RNA-Seq reveals unannotated transcripts and isoform switching during cell differentiation. *Nat. Biotechnol.* 28(5):511–515.

III

Regression

Of the major topics in machine learning and statistical modeling, regression is probably the one that needs least introduction. Regression is concerned with finding quantitative relationships between two or more sets of observations. For example, regression is a powerful way to test for and model the relationship between genotype and phenotype, still one of the most fundamental problems in modern biology. In this chapter, we'll also see how regression (and some of its extensions) can be used to explore the relationship between mRNA and protein abundance at the genome scale.

Regression was probably the first statistical technique: Gauss was trying to fit the errors in regression when he proposed his eponymous distribution, although the name regression was apparently coined by the nineteenth-century geneticist Francis Galton. As we have done in previous chapters, we'll first spend considerable effort reviewing the traditional regression that is probably at least somewhat familiar to most readers. However, we will see that with the advent of numerical methods for local regression, generalized linear models, and regularization, regression is now among the most powerful and flexible of machine learning methods.

III

CHAPTER 7

Univariate Regression

7.1 SIMPLE LINEAR REGRESSION AS A PROBABILISTIC MODEL

Regression aims to model the statistical dependence between i.i.d. observations in two (or more) dimensions, say, $X = X_1, X_2,..., X_n$ and $Y = Y_1, Y_2,..., Y_n$. Regression says that $P(X, Y)$ is not equal to $P(X)P(Y)$, but that we can understand the dependence of Y on X by making a model of how the expectation of Y depends on X. At its simplest, what linear regression says is $E[Y|X] = b_0 + b_1 X$. Notice that this is a *predictive* model: for each value of X, we have a prediction about what we expect Y to be. If we assume that $P(Y|X)$ is Gaussian, then the expectation is just equal to the mean (i.e., $E[Y|X] = \mu$), and we can write the following likelihood function

$$L = P(Y|X, \theta) = \prod_{i=1}^{n} N(Y_i|\mu(X_i), \sigma) = \prod_{i=1}^{n} N(Y_i|b_0 + b_1 X_i, \sigma)$$

$$= \prod_{i=1}^{n} \frac{1}{\sigma\sqrt{2\pi}} e^{-((Y_i - b_0 - b_1 X_i)^2)/2\sigma^2}$$

where the product is over the i.i.d. observations, and I've written $N()$ to represent the Gaussian distribution where the mean for each observation of Y depends on the corresponding observation X. As with every probabilistic model, the next step is to estimate the parameters, and here, these are b_0, b_1, and s. Figure 7.1 illustrates the probabilistic interpretation of simple linear regression. In the case of simple linear regression, it is possible to differentiate (the log of) this objective function with respect to the parameters to obtain closed forms for the maximum likelihood estimators.

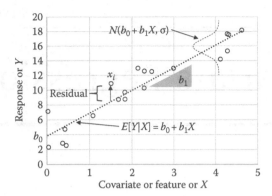

FIGURE 7.1 The probabilistic interpretation of simple linear regression. Parameters of the model are b_0, b_1, and σ. Circles represent observations of X and Y. Only the response (Y) variable is assumed to have "noise" in this model.

Before we proceed to find formulas for the maximum likelihood estimators (MLEs), however, I think it's useful to consider this likelihood function a bit further. First, compare it to the case where we simply model Y using a single Gaussian: that turns out to be the case of $b_1 = 0$. This is the first attractive feature of simple linear regression: when there is no relationship between X and Y (at least no relationship that can be captured by the linear model), linear regression becomes a simple Gaussian model for the variable whose values we are trying to predict.

Another important point is to consider the presence of randomness implied by the model: when we say that the expected value of Y depends on X through a simple formula, we are saying that although we expect Y to be connected to X, we accept that our observations of Y can have some randomness associated with them (which in the model is assumed to be predicted by the Gaussian distribution.) However, there is no place in this likelihood where we include the possibility of randomness in X: we are assuming that X is, in some sense, "perfectly" measured, and all of the randomness is our observations of Y. This implies that linear regression is not symmetric: if we write the model $L = P(Y|X, \theta)$, we will get a different answer than if we write $L = P(X|Y, \theta)$.

7.2 DERIVING THE MLEs FOR LINEAR REGRESSION

Let's now go ahead and derive the MLEs for some of the parameters in the simple linear regression model. As usual, we first take the log of the likelihood. We have

$$\log L = \log \prod_{i=1}^{n} \frac{1}{\sigma\sqrt{2\pi}} e^{-((Y_i - b_0 - b_1 X_i)^2)/2\sigma^2} = \sum_{i=1}^{n} -\log \sigma\sqrt{2\pi} - \frac{(Y_i - b_0 - b_1 X_i)^2}{2\sigma^2}$$

We next take the derivatives with respect to each of the parameters and set these derivatives to zero. For example,

$$\frac{\partial \log L}{\partial b_0} = \sum_{i=1}^{n} -0 + 2\frac{(Y_i - b_0 - b_1 X_i)}{2\sigma^2} = \sum_{i=1}^{n} \frac{(Y_i - b_0 - b_1 X_i)}{\sigma^2} = 0$$

where the first term is zero because it did not depend on b_0, and the derivative of the squared term is just -1. Notice that since σ^2 doesn't depend on i, we can take it out of the sum and multiply both sides of the equation by σ^2.

$$-\sum_{i=1}^{n} b_0 + \sum_{i=1}^{n} (Y_i - b_1 X_i) = 0$$

Finally, the sum of b_0 from $i = 1$ to n is simply n times b_0.

$$-nb_0 + \sum_{i=1}^{n} (Y_i - b_1 X_i) = 0$$

We can now solve for b_0 to get

$$b_0 = \frac{1}{n}\left(\sum_{i=1}^{n} (Y_i - b_1 X_i) \right) = \frac{1}{n}\left(\sum_{i=1}^{n} Y_i - b_1 \sum_{i=1}^{n} X_i \right) = \frac{1}{n}\sum_{i=1}^{n} Y_i - b_1 \frac{1}{n}\sum_{i=1}^{n} X_i$$

where I've done a bit of rearranging for clarity. Consider this equation in the context of the "null hypothesis" for linear regression, namely that $b_1 = 0$. This equation says that under the null hypothesis, b_0 is just the average of Y, which, as we have seen in Chapter 4, turns out to be the MLE for the μ parameter of the Gaussian distribution. This makes sense based on what I already said about the regression model.

Notice that as often happens, the MLE for b_0 depends on b_1, so that to maximize the likelihood, we will have to simultaneously solve the equation

$$\frac{\partial \log L}{\partial b_1} = \sum_{i=1}^{n} -0 + 2\frac{(Y_i - b_0 - b_1 X_i)}{2\sigma^2} X_i = \sum_{i=1}^{n} \frac{(Y_i - b_0 - b_1 X_i)}{\sigma^2} X_i = 0$$

where the first term is zero because it did not depend on b_1, and the derivative of the squared term is $-X_i$. Once again, we can take σ^2 out of the sum and multiply both sides of the equation by σ^2, leaving

$$\sum_{i=1}^{n}(Y_i - b_0 - b_1 X_i)X_i = \sum_{i=1}^{n} Y_i X_i - \sum_{i=1}^{n} b_0 X_i - \sum_{i=1}^{n} b_1 X_i^2 = 0$$

We can solve for b_1

$$b_1 = \frac{\sum_{i=1}^{n} Y_i X_i - b_0 \sum_{i=1}^{n} X_i}{\sum_{i=1}^{n} X_i^2}$$

Luckily, the equation for b_1 only depends on b_0, so that we have two equations and two unknowns and we can solve for both b_0 and b_1. To do so, let's plug the equation for b_0 into the equation for b_1.

$$b_1 = \frac{\sum_{i=1}^{n} Y_i X_i - (1/n)\left(\sum_{i=1}^{n} Y_i\right)\left(\sum_{i=1}^{n} X_i\right) + b_1(1/n)\left(\sum_{i=1}^{n} X_i\right)\left(\sum_{i=1}^{n} X_i\right)}{\sum_{i=1}^{n} X_i^2}$$

After algebra, I have

$$b_1 \frac{\sum_{i=1}^{n} X_i^2 - (1/n)\left(\sum_{i=1}^{n} X_i\right)\left(\sum_{i=1}^{n} X_i\right)}{\sum_{i=1}^{n} X_i^2}$$

$$= \frac{\sum_{i=1}^{n} Y_i X_i - (1/n)\left(\sum_{i=1}^{n} Y_i\right)\left(\sum_{i=1}^{n} X_i\right)}{\sum_{i=1}^{n} X_i^2}$$

which can be solved to give

$$b_1 = \frac{\sum_{i=1}^{n} Y_i X_i - (1/n)\left(\sum_{i=1}^{n} Y_i\right)\left(\sum_{i=1}^{n} X_i\right)}{\sum_{i=1}^{n} X_i^2 - (1/n)\left(\sum_{i=1}^{n} X_i\right)\left(\sum_{i=1}^{n} X_i\right)} = \frac{m_{XY} - m_X m_Y}{m_{X^2} - m_X^2}$$

where, to avoid writing out all the sums, it's customary to divide the top and bottom by n, and then put in m_X to represent various types of averages. This also works out to

$$b_1 = \frac{m_{XY} + m_X m_Y}{s_X^2} = r \frac{s_Y}{s_X}$$

where

s are the standard deviations of each list of observations

r is the "Pearson's correlation" $r(X,Y) = \left(\sum_{i=1}^{n} (X_i - m_X)(Y_i - m_Y) \right) / s_X s_Y$

This equation for b_1 is interesting because it shows that if there is no correlation between X and Y, the slope (b_1) must be zero. It also shows that if the standard deviation of X and Y are the same ($s_Y/s_X = 1$), then the slope is simply equal to the correlation.

We can plug this back in to the equation for b_0 to get the second MLE:

$$b_0 = m_Y - r \frac{s_Y}{s_X} m_X$$

7.3 HYPOTHESIS TESTING IN LINEAR REGRESSION

Because the parameters for the linear regression model have been estimated by maximum likelihood, we can use the asymptotic results about MLEs to tell us the distribution of the parameter estimates. We know that the estimates will be asymptotically Gaussian with mean equal to the true value of the parameter and the variance related to the second derivative of the likelihood evaluated at the maximum. Once again, this can all be computed analytically so that the distribution of the parameters is known to be $N(b_1, (\sigma/\sqrt{n} s_X))$. Combined with the convenient null hypothesis that $b_1 = 0$, we can compare $N(b_1, (\sigma/\sqrt{n} s_X))$ to 0 to put a P-value on any regression model. Furthermore, we can formally compare whether the slopes of two different regression models are different. In practice, the variance of the MLEs converges asymptotically, so it works well only when the number of datapoints is large (e.g., Figure 7.2) Of course, there are also approximations for small samples sizes that are implemented in any standard statistics software.

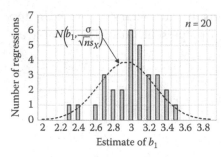
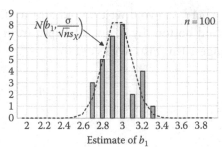

FIGURE 7.2 Convergence of the MLEs to the Gaussian distributions predicted by statistical theory. In each case, 30 regressions were calculated on randomly generated data where the true slope was equal to 3. In the left panel, regressions were fit based on 20 datapoints, while in the right panel, 100 datapoints were used.

DERIVING THE DISTRIBUTION OF THE MLE FOR b_1

It's a useful exercise to do the calculations to obtain the variance of the MLE for b_1. As I mentioned in Chapter 4, we need the matrix of the second derivatives of the likelihood, evaluated at the maximum of the likelihood. The matrix of second derivatives turns out to be

$$
FI = \begin{vmatrix}
\dfrac{\partial^2 \log L}{\partial b_0^2} & \dfrac{\partial^2 \log L}{\partial b_1 \partial b_0} & \dfrac{\partial^2 \log L}{\partial b_0 \partial \sigma} \\[2mm]
\dfrac{\partial^2 \log L}{\partial b_0 \partial b_1} & \dfrac{\partial^2 \log L}{\partial b_1^2} & \dfrac{\partial^2 \log L}{\partial \sigma \partial b_1} \\[2mm]
\dfrac{\partial^2 \log L}{\partial b_0 \partial \sigma} & \dfrac{\partial^2 \log L}{\partial b_1 \partial \sigma} & \dfrac{\partial^2 \log L}{\partial \sigma^2}
\end{vmatrix}
= -\frac{1}{\sigma^2}
\begin{bmatrix}
n & \sum_{i=1}^{n} X_i & 0 \\[2mm]
\sum_{i=1}^{n} X_i & \sum_{i=1}^{n} X_i^2 & 0 \\[2mm]
0 & 0 & 2n
\end{bmatrix}
$$

To show you how I got all those second derivatives, we will need the derivatives of the likelihood with respect to σ and the MLE for this parameter.

$$
\frac{\partial \log L}{\partial \sigma} = \sum_{i=1}^{n} -\frac{1}{\sigma} + 2\frac{(Y_i - b_0 - b_1 X_i)^2}{2\sigma^3} = \frac{1}{\sigma} \sum_{i=1}^{n} -1 + \frac{(Y_i - b_0 - b_1 X_i)^2}{\sigma^2}
$$

Setting this to zero and solving will give us the MLE for the "noise" parameter in the linear regression

$$
\frac{n}{\sigma} = \frac{1}{\sigma} \sum_{i=1}^{n} \frac{(Y_i - b_0 - b_1 X_i)^2}{\sigma^2}
$$

$$\sigma^2 = \frac{1}{n} \sum_{i=1}^{n} (Y_i - b_0 - b_1 X_i)^2$$

To get the partial derivatives, the order doesn't matter, so we can do either derivative first (the matrix will be symmetric). Using the formula for the derivative of the likelihood with respect to b_0 from above, we can we take the second derivative with respect to σ as follows:

$$\frac{\partial^2 \log L}{\partial b_0 \partial \sigma} = \frac{\partial}{\partial \sigma} \sum_{i=1}^{n} \frac{(Y_i - b_0 - b_1 X_i)}{\sigma^2} = \frac{-2}{\sigma^3} \sum_{i=1}^{n} (Y_i - b_0 - b_1 X_i) = -\frac{2}{\sigma} \left(\frac{\partial \log L}{\partial b_0} \right) = -\frac{2}{\sigma}(0) = 0$$

In the last two steps, I did something very clever: I wrote the second derivative in terms of the first derivative that we calculated. Because the distribution of the MLEs is related to the matrix of second derivatives, *at the maximum of the likelihood*, we know that the first partial derivative must be zero—that's how we know we're at the maximum—and that's how we derived the MLEs. You can see that the same story will be true for the partial derivatives with respect to σ and b_1 as well.

$$\frac{\partial^2 \log L}{\partial b_1 \partial \sigma} = \frac{\partial}{\partial \sigma} \sum_{i=1}^{n} \frac{(Y_i - b_0 - b_1 X_i)}{\sigma^2} X_i = -\frac{2}{\sigma} \left(\frac{\partial \log L}{\partial b_1} \right) = -\frac{2}{\sigma}(0) = 0$$

Not all the second derivatives turn out to be zero. For example, here's the second derivative of the likelihood with respect to b_1:

$$\frac{\partial^2 \log L}{\partial b_1^2} = \frac{\partial}{\partial b_1} \left[\sum_{i=1}^{n} \frac{(Y_i - b_0 - b_1 X_i)}{\sigma^2} X_i \right] = -\sum_{i=1}^{n} \frac{X_i^2}{\sigma^2}$$

where again I used the formula we already had for the first derivative. A little more tricky is the second derivative with respect to the noise parameter:

$$\frac{\partial^2 \log L}{\partial \sigma^2} = \frac{\partial}{\partial \sigma} \left[-\frac{n}{\sigma} + \sum_{i=1}^{n} \frac{(Y_i - b_0 - b_1 X_i)^2}{\sigma^3} \right] = \frac{n}{\sigma^2} - \frac{3}{\sigma^2} \sum_{i=1}^{n} \frac{(Y_i - b_0 - b_1 X_i)^2}{\sigma^2}$$

Although this looks bad, remember that at the maximum, $(\partial \log L)/(\partial \sigma) = 0$, and we already found that this means that

$$\frac{n}{\sigma} = \frac{1}{\sigma} \sum_{i=1}^{n} \frac{(Y_i - b_0 - b_1 X_i)^2}{\sigma^2}$$

Therefore, $\sum_{i=1}^{n} ((Y_i - b_0 - b_1 X_i)^2 / \sigma^2)$ must actually be just n. We have

$$\frac{\partial^2 \log L}{\partial \sigma^2} = \frac{n}{\sigma^2} - \frac{3}{\sigma^2} n = -\frac{2n}{\sigma^2}$$

Putting all second derivatives together, we get the matrix shown earlier. But we still need the inverse of this matrix, which depends on the determinant. The determinant of a 3×3 matrix is a messy thing, but because of the all the zeros, it's reasonable to calculate it in this case.

$$FI^{-1} = \frac{-\sigma^2}{2nn\sum_{i=1}^{n} X_i^2 - 2n\left(\sum_{i=1}^{n} X_i\right)^2} \begin{bmatrix} 2n\sum_{i=1}^{n} X_i^2 & -2n\sum_{i=1}^{n} X_i & 0 \\ -2n\sum_{i=1}^{n} X_i & 2nn & 0 \\ 0 & 0 & n\sum_{i=1}^{n} X_i^2 - \left(\sum_{i=1}^{n} X_i\right)^2 \end{bmatrix}$$

I left some spaces to show where the various terms in the inverse came from. The inverse can be simplified by factoring out $2n$ and rewriting in terms of s_X^2, the variance of the data X.

$$FI^{-1} = \frac{-\sigma^2}{n^2 s_X^2} \begin{bmatrix} \sum_{i=1}^{n} X_i^2 & -\sum_{i=1}^{n} X_i & 0 \\ -\sum_{i=1}^{n} X_i & n & 0 \\ 0 & 0 & \frac{n s_X^2}{2} \end{bmatrix}$$

Finally, the variance of the MLE for b_1 is the expectation of the negative of the middle entry in the matrix $E[-(FI^{-1})_{22}] = \sigma^2 / (n s_X^2)$. Taking the expectation has no effect in this case because there are no random variables left in our formula (the expectation of the variance is just the variance, and the expectation of our estimator is just the estimator). So the distribution of the MLE is a Gaussian with the mean parameter equal to b_1 and the standard deviation equal to $\sigma / \sqrt{n} s_X$. Notice that the standard deviation of the MLE decreases proportional to the square root of the number of datapoints n, and that the variance depends on the MLE for the noise parameter: the further the data are from the regression line, the larger the variance of our estimate of the slope. Another interesting point here is that the off-diagonal terms of the matrix in the rows for b_0, and b_1 are not zero. This means that the joint distribution of b_0 and b_1 is a multivariate Gaussian with nondiagonal covariance. Although in our model the parameters are independent, our *estimates* of the parameters are not.

Even more powerful than using the null hypothesis, $b_1 = 0$, is using the related null hypothesis $r = 0$. Although the formula for the correlation is a bit complicated, one important thing to notice is that it is symmetrical in X and Y. So although regression in general is not symmetric in X and Y, the symmetrical part of the relationship between X and Y is captured by the correlation. This is one of the reasons that the correlation is a very useful distance measured between two vectors (as discussed in Chapter 5.) Perhaps even better is that the distribution of the correlation is known (approximately) under the null hypothesis of no association between X and Y. Assuming that X and Y are truly Gaussian, but are independent, then the statistic

$$t = r\sqrt{\frac{n-2}{1-r^2}}$$

where n is the number of observations, has a t-distribution with $n - 2$ degrees of freedom. This means that, given two lists of numbers, X and Y, you can go ahead and test whether there is an association between them without having to assume anything about which one causes the other.

It's important to remember that the P-value approximation for the correlation and the asymptotic normality of the estimate of b_1 *do* assume that the data are at least approximately Gaussian. However, in the context of hypothesis testing, there is a beautiful way around this: replace the actual values of X and Y with their ranks and compute the Pearson correlation on the ranks. It turns out that the distribution of this correlation (known as the "Spearman correlation") under the null hypothesis is also known! The Spearman correlation is widely used as a nonparametric test for association without any assumptions about the distribution of the underlying data.

As with the rank-based tests we discussed in Chapter 2, dealing with tied ranks turns out to make the formulas for the Spearman correlation a bit complicated in practice. Nevertheless, this test is implemented correctly in any respectable statistics software.

7.4 LEAST SQUARES INTERPRETATION OF LINEAR REGRESSION

Although I have presented linear regression in the context of a probabilistic model, where the expectation of the Gaussian distribution was assumed to depend on X, notice that the log-likelihood can also be written as

$$\log L = \sum_{i=1}^{n} -\log \sigma \sqrt{2\pi} - \frac{(Y_i - b_0 - b_1 X_i)^2}{2\sigma^2}$$

$$= -n \log \sigma \sqrt{2\pi} - \frac{1}{2\sigma^2} \sum_{i=1}^{n} (Y_i - b_0 - b_1 X_i)^2$$

Maximizing the likelihood function with respect to b_1 and b_0 is equivalent to minimizing $\sum_{i=1}^{n} (Y_i - b_0 - b_1 X_i)^2$. The term that is being squared corresponds to the difference between the observed Y and the expected or predicted Y (which is just $b_0 + b_1 X$). This difference is also called the "residual" for that observation—the part of Y that was not successfully explained by X. Maximizing the likelihood is therefore equivalent to minimizing the sum of squared residuals or the SSR, where

$$SSR = \sum_{i=1}^{n} (Y_i - b_0 - b_1 X_i)^2$$

Therefore, even if you don't believe that your data are Guassian at all, you can still go ahead and fit a standard regression model to find the "ordinary least squares" (OLS) estimators, which minimize the squared differences between the observations and the predictions. In the case of simple linear regression, the OLS estimators turn out to be the same as the ML estimators if the data are assumed to be Gaussian.

To evaluate the fit of the regression model in this context, we might be interested in asking: how much of the Ys were successfully explained by X? To make this a fair comparison, we should standardize the SSR by the total amount of variability that there was in Y to begin with: if Y never changed at all, it's not fair to expect X to explain much at all. The standard way to do this is to compare the SSR to the standard deviation of Y using the r^2

$$r^2 = \frac{s_Y^2 - (SSR/n)}{s_Y^2} = 1 - \frac{SSR}{n s_Y^2} = 1 - \frac{\sum_{i=1}^{n} (Y_i - E[Y_i|X_i])^2}{\sum_{i=1}^{n} (Y_i - E[Y])^2}$$

If the predictions of Y based on X were perfect, the residuals would all be zero, and the r^2 (also known as R^2) would be 1. The r^2 can be thought of as the fraction of the variance of Y that is still present in the residuals. In the context of the "null hypothesis" of $b_1 = 0$, this means that X didn't predict anything about Y, and the regression was equivalent to estimating a simple Gaussian model for Y. Said another way, the r^2 measures the amount of Y that can actually be explained by X, or the differences between $Y|X$ and Y on its own. Importantly, this interpretation is valid regardless of the underlying distributions of X and Y. Conveniently, r^2 is also very simple to calculate: Just take the correlation and square it! So even if your data are not Gaussian, you can still say something meaningful about how much of the variance of Y is explained by its relationship to X.

7.5 APPLICATION OF LINEAR REGRESSION TO eQTLs

To illustrate the power of simple linear regression for hypothesis testing in modern molecular biology, let's consider a very difficult statistical problem: detecting eQTLs. In Chapter 3, we saw that eQTLs can be identified using a simple t-test that compares the distribution of expression levels for one genotype to the other. The null hypothesis is that the two genotypes have the same expression, so we find eQTLs when we reject it. (We also saw that this leads to a formidable multiple testing problem, with which simple regression doesn't help us; but see the discussion in Chapter 9.) Here, I just want to show an example of how regression can be used as a test of association in place of the *t*-test we did before. When we formulate the search for eQTLs as a regression problem, we're regressing a quantitative measure of gene expression (Y) on the genotypes of each individual (X). Although the genotypes at each locus are discrete (AA, Aa, aa in diploids or A and a in haploids), this isn't actually a technical problem for linear regression because the Gaussian model applies only to the Y variable; the Xs are assumed to be perfectly observed.

The main advantage of regression in this case is interpretability: when we calculate the correlation between the phenotype (the gene expression level) and the genotype at each locus, the R^2 summarizes how much of the variation in the expression level is actually explained by the genotype. In addition to the biological interpretation of the test based on R^2, regression also turns out to be more flexible—it will generalize more easily to multiple dimensions and more easily accommodate differences in ploidy in the experiment.

For example, although it might not look like the typical illustration of linear regression shown earlier, the expression level of *AMN1* is strongly correlated with the genotype at a marker on chromosome 2 in yeast. The correlation, r, is −0.847, but note that the sign (positive or negative) is arbitrary: We assigned the reference genotype to be 0 and the mutant to be 1. We could have done it the other way and gotten a positive correlation. The associated t-statistic is −16.2714, with $df = 104$, which, as you can imagine, is astronomically unlikely to occur under the null hypothesis where the correlation is actually 0 (the P-value is less than 10^{-10}). In addition, we can go ahead and square the correlation, which gives $R^2 = 0.718$, suggesting that the genotype at this locus explains almost three quarters of the variance in *AMN1* expression. Figure 7.3a shows the data.

We can go ahead and perform tests like this systematically to identify eQTLs in the data. I chose 10,000 random pairs of genes and markers and computed the correlation coefficients and associated P-values. In Figure 7.3b, you can see that the distribution of P-values has an excess of small P-values indicating that there are probably a large number of significant associations. To see what the typical effect size is, I plotted the R^2 for all the tests that were significant at 0.05 after a Bonferroni correction (bottom panel in

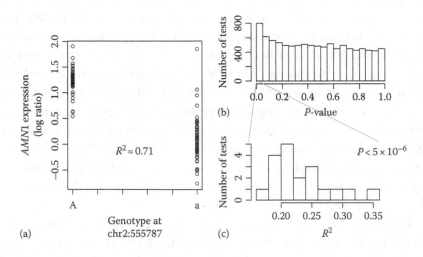

FIGURE 7.3 A simple regression of gene expression levels on genotype. (a) The dependence of the relative gene expression level of *AMN1* in an yeast cross on genotype. Each circle represents a segregant of the cross. (b) The distribution of P-values obtained from tests of association between genotype and gene expression level. (c) R^2 values associated with tests that passed the Bonferroni multiple testing correction. Typical R^2 values are much less than what is seen for *AMN1*.

Figure 7.3b). Even though these associations are very strongly significant (remember that the Bonferroni correction is likely to be very conservative), the typical effect sizes are much less than what we saw for AMN1. For these examples, the genotype explains more like 20% or 25% of the variation in gene expression levels. Thus, although tests for eQTLs can be performed using a variety of statistical techniques (*t*-tests, WMW, etc.), summarizing the statistical association using the fraction of variance explained gives us the insight that *AMN1* is probably quite unusual with respect to how much of its gene expression variation is controlled by a marker at a single locus, even among the strongest eQTLs in the experiment.

7.6 FROM HYPOTHESIS TESTING TO STATISTICAL MODELING: PREDICTING PROTEIN LEVEL BASED ON mRNA LEVEL

By far, the most common use of linear regression in molecular biology is to detect statistical associations as illustrated in the eQTL example—to reject the hypothesis that X and Y are independent (by calculating the P-value associated with a correlation coefficient). This is because we don't usually have quantitative predictions about the relationships between biological variables—we are looking for differences between mutant and wt, but we don't have a particular hypothesis about what that difference should be. Or we know that evolutionary rates should be slower for more important proteins, but we don't really know how much slower.

However, regression can also be used to quantify the relationship between variables that we *know* are correlated. For example, the central dogma of molecular biology is: DNA makes RNA makes protein. Let's try to turn this into a quantitative prediction (following Csárdi et al. 2015): under the assumptions that (1) RNA and protein degradation are constant, and (2) the cell has reached some kind of equilibrium state, the abundance of each protein in the cell would simply be proportional to the mRNA for that protein, with the constant given by the ratio of translation to degradation rates.

$$Protein = \frac{Translation\ rate}{Degradation\ rate} mRNA$$

In this case, we aren't really interested in rejecting the null hypothesis of independence—we know proteins depend on mRNA—instead, we want to test if there really is a simple *linear* relationship between protein abundance and mRNA abundance. If we take logs of both sides, we get

$$\log Protein = \log \frac{Translation\ rate}{Degradation\ rate} + \log mRNA$$

Setting $Y = \log Protein$ and $X = \log mRNA$, we have a regression problem

$$E[Y|X] = \log \frac{Translation\ rate}{Degradation\ rate} + X$$

where b_0 is the log of the translation to degradation ratio, and if there really is a simple linear relationship between protein and mRNA levels, we expect to find $b_1 = 1$. This means that we can use this simple regression to estimate b_1, and because we know our estimate of b_1 will be Gaussian with standard deviation given, we can test it is statistically different from 1, thus testing the hypothesis that there is a simple linear relationship between mRNA and protein levels.

The results of this analysis are shown in Figure 7.4. If we do the analysis on the whole dataset, we obtain a slope (b_1) of 0.837 and the standard deviation of the estimate is less than 0.02. This means that we are more than 8 standard deviations away from 1. Since we know the distribution of b_1 should be Gaussian, this is an incredibly unlikely observation if the slope was truly 1. However, although we can rule out 1 with large statistical confidence, 0.837 is actually not that far from 1.

Looking at the data more carefully (see Figure 7.4), it's clear that something funky is happening with our data at high mRNA levels ($\log mRNA > 3$). Instead of protein levels rising simply, there seem to be two classes of genes: those with high protein levels as expected (around log Protein = 10) and those with unexpectedly low protein levels (around log Protein = 8). These genes are violating not only the simple linear assumption, but even the assumption that there is any consistent relationship between mRNA and protein levels. Another way to see this is to look at the distribution of residuals. As I mentioned, the residual is the difference between the observed Ys and the predicted Y, that is, $Y_i - E[Y|X_i]$. The distributions of residuals are shown at the bottom panel of the figure for the genes with typical mRNA levels and for genes with the most abundant transcripts. It's clear from these histograms that there is a subpopulation of highly expressed genes for which the model is predicting much bigger protein levels than are observed (bump in the histogram around −5). It turns out that these highly expressed genes represent less than 5% of the data in the analysis.

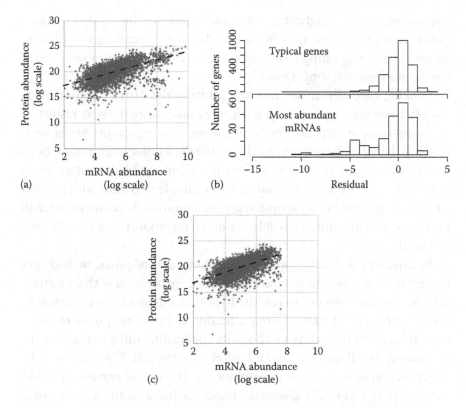

FIGURE 7.4 A regression model of the relationship between mRNA and protein abundance. Each gray symbol represents one of >4000 yeast genes for which at least three mRNA and protein measurements were available. (Data were collated by Csárdi, G. et al., 2015.) (a) The regression model for the whole dataset, (b) the residuals of the regression with genes divided into two categories based on their abundance, and (c) the regression after removing the genes with the largest mRNA abundances.

If we remove these genes with high abundance transcripts, we obtain the analysis shown on the bottom of the figure. Doing the regression analysis on these genes gives an estimate of the slope (b_1) to be 0.997 with standard deviation of 0.022. Needless to say, this is *very* close to the slope of 1 that we predicted, and we can say that the data is consistent with our model—in fact, we can say that for this 95% of the data, the simple linear model explains the data to about 2% accuracy. The regression analysis indicates that on average, there probably is something close to a linear relationship between protein and mRNA abundance for most genes, but that this relationship breaks down for the most highly expressed genes in the genome. Interestingly, if we continue to remove the highly expressed genes from the dataset, the estimate of the slope actually gets a little bit larger than 1, suggesting that even on average there are probably subtle deviations from the simple model.

Looking carefully at the data and plotting the distribution of residuals is a powerful general approach to evaluating regression results. If you want to do statistical hypothesis testing on a regression model, confirming that the distribution of the residuals follows your assumptions is the best way to know if your P-values (and confidence intervals on parameters) are likely to be reliable. This example also illustrates another important general point: simple regression analysis is sensitive to small numbers of datapoints (possibly outliers) disproportionately affecting the result.

Because in this analysis we included thousands of genes, we had very large statistical power to test our hypothesis. However, as with all statistical modeling, our ability to test the hypothesis depends on the assumptions. In this case, I want to draw attention to the assumptions of linear regression about noise. It is a technically challenging thing to measure the abundance for all the proteins and mRNAs in the cell. There is sure to be experimental noise in these measurements. The simple regression model we used here included a Gaussian model for noise in the log (so-called "multiplicative noise," see Exercises) of the protein measurements, Y. However, the model we used assumes that the mRNA levels, X, are perfectly measured, which is almost certainly not right. It turns out that there are regression techniques (beyond the scope of this book) that allow noise in both the Y and X to be included in the model. In this case, that type of model turns out be more appropriate, and the story is more complicated that what I described here (Csárdi et al. 2015).

FIVE GREAT THINGS ABOUT SIMPLE LINEAR REGRESSION

- Fast, simple formulas to estimate parameters
- Straightforward hypothesis testing in the maximum likelihood framework
- Interpretable measures of model fit even when the distribution of the data is unknown
- Nonparametric tests of association based on Spearman's rank correlation
- Easy to evaluate the assumptions using the distribution of residuals

7.7 REGRESSION IS NOT JUST "LINEAR"—POLYNOMIAL AND LOCAL REGRESSIONS

One very powerful feature of linear regression is that it can be generalized in several different ways. First, it does not assume that the relationship between variables is "linear" in the sense that there has to be a line connecting the dots. A simple way to extend linear regression beyond a simple line is to think about regressions on transformations of the X variables (features or covariates). For example, it's totally okay to write regressions like

$$E[Y|X] = b_0 + b_1 X^2$$

$$E[Y|X] = b_0 + b_1 \log X$$

$$E[Y|X] = b_0 + b_1 X + b_2 X^2$$

Because linear regression assumes the Xs are measured exactly, everything we worked out will apply just fine for these transformations of X. You might notice that in one of the regressions I added an extra parameter to weight the term corresponding to X^2. Although at first it looks like this will make the regression problem much more complicated, if you write out the likelihood for this regression (see Exercises), you'll see that you'll be able to derive a third formula for b_2, just like we derived formulas for b_0 and b_1. Although the algebra gets a bit tedious, in Chapter 8, we will discuss how to write out the equations for linear regression in linear algebra notation and solve them for an arbitrary number of b's. Thus, "linear" regression can be used on data where arbitrary nonlinear transformations have been applied to the predictor. In the case where we fit a term proportional to X and X^2, we are using linear regression to fit a parabola (or quadratic form) to our data. Thus, polynomials of arbitrary degree can be fit using linear regression.

Perhaps even more powerful than transformations of X, are extensions of regression that do not use all of the values of X equally. These methods are known as "local" regressions. The idea is very simple: if you have plenty of data, you can use only nearby points to make a model that predicts well in a local region or neighborhood of the space. The simplest form of a local regression is "nearest neighbor regression" where, instead of the entire dataset, only a fixed window of, say, k, datapoints are used to predict Y. In a nearest neighbor regression, we construct a simple average at each observation X_i. The k-nearest datapoints can be chosen using any of the distance metrics we discussed in Chapter 5. In the univariate (one-dimensional) case, this just means the numerically closest points. I hope it's clear that different subsets of the dataset will have different averages and the overall predictions of Y_i need not form a line at all. As k approaches the size of the entire dataset, the results for nearest neighbor regression will approach the average of the dataset. The main drawback of nearest neighbor regression is that it's very sensitive to the size of the neighborhood, and that we only get predictions of Y at observed values of X_i. For this reason, nearest neighbor regression often includes heuristic methods to interpolate between the X_i.

Perhaps the most elegant formulation of local regression is known as kernel regression. In this formulation, a weighted average is used to predict Y at any point X_0, based on all the data, but using a weighting scheme that weights the Xs so that data nearby X_0 contribute more strongly to the average than distant datapoints. Specifically, in the so-called Nadaraya–Watson kernel regression, the prediction of Y at a point X_0 is given by

$$E[Y|X_0] = \frac{\sum_{i=1}^{n} K(|X_i - X_0|)Y_i}{\sum_{i=1}^{n} K(|X_i - X_0|)}$$

where $K(|X_i - X_0|)$ is now the so-called "kernel function," which is used to weight the data based on their distance to the point of interest $|X_i - X_0|$. Thus, kernel regression solves both the problems of nearest neighbor regression, but raises the question of how to choose the kernel function. We seek a function that is maximal when the distance between points is small and then decays rapidly when the distance is large. A great example of such a function is the Gaussian probability density, and this function is a very popular choice for a kernel (where it is known as the radial basis function kernel or RBF kernel).

When we use a Gaussian kernel, we also need to choose the standard deviation (or bandwidth) of the kernel. This determines how fast the kernel decays as a function of distance, effectively setting the size of the neighborhood of points used for local regression. If the bandwidth (the standard deviation of the Gaussian kernel) is too small, nearby points will dominate the local estimate, and the estimate will be too noisy. On the other hand, if the bandwidth is too large, the local estimate will be insensitive to the variation in the data. In the limit of infinite bandwidth, kernel regression converges to a simple average of the data (see Exercises). There are methods to choose this bandwidth automatically based on the data, but these should be used with caution. In most cases, you can obtain a good (low-dimensional) kernel regression by trying a few values of the bandwidth.

To illustrate the effect of the bandwidth on kernel regression, I fit two kernel regressions to the mRNA and protein data (Figure 7.5). Notice that the result confirms that over a large range of mRNA and protein, the relationship appears linear; but at high mRNA levels, the linear relationship clearly becomes more complicated. However, when we fit the regression with a bandwidth that is too small, we find that the kernel regression becomes unstable and follows extreme datapoints.

The most widely used form of local regression is LOESS (Cleveland and Delvin 1988). In this method, a polynomial is used to model the conditional expectation of Y any point X_0, based on a weighting scheme that weights the Xs, so that only the nearby X_0 is given positive weights, and everything else is given a very small weight or weight of zero. Note that as with kernel regression, X_0 is not necessarily an actual datapoint (or observation)—it can be, but it can also just be a point where we want an estimate of Y. Just as with standard linear regression we obtain a prediction at every value

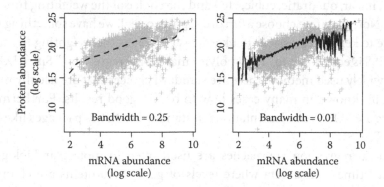

FIGURE 7.5 Kernel regression of protein levels on mRNA levels. On the left is a kernel regression with a suitable choice of bandwith (0.25) for this data. On the right is a kernel regression with a very small bandwidth (0.01) that clearly shows overfitting of the data. Gray "+"s indicate individual genes for which protein and mRNA abundance have been measured.

of X, local regression gives a prediction at every value of X based on the nearby data.

The estimate at X_0 is obtained by first fitting a polynomial to the nearby points using linear regression by minimizing a weighted version of the SSR.

$$SSR = \sum_{i=1}^{n} w(|X_i - X_0|)\left(Y_i - b_0 - b_1 X_i + b_2 X_i^2\right)^2$$

where I have written the weighting function w as a function of $|X_i - X_0|$ to emphasize that it is a function of the distance of the datapoint from the point of interest. Because these weights don't depend on the parameters of the regression, they add no numerical complication to the minimization of the SSR (see Exercises). Once we have obtained the local estimates of the b parameters, we can go ahead and predict the value of Y at the point X_0.

$$E[Y|X_0] = b_0 + b_1 X + b_2 X^2$$

Amazingly, many of the results from standard linear regression, regarding the convergence of the predictions of Y and the variance in the estimate, can be generalized to weighted regressions like kernel regression and LOESS. Of course, two important issues are the choice of the polynomial (e.g., linear, quadratic, cubic, etc.) and the choice of the weighting function $w()$. Note that if we choose a 0th order polynomial, we have something very close to kernel regression. If a kernel is chosen as the weighting function, LOESS is equivalent to "local polynomial kernel regression." Since LOESS is a widely used method, there are standard choices that are usually made, and it's known in many cases how to obtain good results. Furthermore, there are LOESS implementations in standard statistics packages like R or MATLAB®.

Local regression approaches are used often in molecular biology to model time-course data where levels of genes or proteins are changing over time, but there is no reason to assume that these changes are linear functions (or any other simple form) of time.

Another interesting application of local polynomial regressions is that they can be used to estimate derivatives of the curves that they fit. For example, in the mRNA and protein modeling example, we wanted to test if the slope of the (log-transformed) regression was 1, but we noticed that

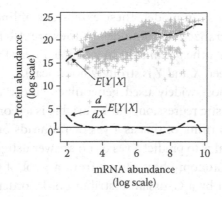

FIGURE 7.6 A second-order kernel regression (with bandwidth 0.5) is used to model the conditional expectation $E[Y|X]$ (upper dashed trace), and the first derivative $(d/dX)E[Y|X]$ (lower dashed trace). The hypothesis of simple proportionality of mRNA and protein abundances is indicated as a solid line (1) and the derivative follows this prediction over the range where most of the data is found.

the assumption of a constant slope over the whole data was unrealistic. I solved this by simply throwing out the data (at high mRNA levels) that seemed to violate the simple assumptions about the relationship between mRNA and protein abundance. However, using a polynomial kernel regression, we can estimate the slope (the first derivative) of the regression at every point without assuming a simple linear relationship in the data. A kernel estimate of the derivative is shown in Figure 7.6: you can see that for most of the range of mRNA concentrations, the slope is very close to 1. Note that unlike in the simple linear regression models shown in Figure 7.3, no assumptions about a simple "line" fitting the whole dataset are made in order to estimate the slope here. And still, we find that on average, there is a simple linear relationship between mRNA and protein abundance for the range of expression levels where most of the mRNAs are found.

7.8 GENERALIZED LINEAR MODELS

Linear regression is even more general than the transformations of X listed and the weighted "local" regressions that are widely used to fit data with nonlinear dependence on X. For a large class of functions of X (called "link functions" in R), it's also possible to use the same mathematical machinery as in linear regression, but to relax the assumption that Y is a real number (between $-\infty$ and ∞) whose probability distribution is described by a Gaussian model. These extensions of linear regression are referred to as generalized linear models as they are still linear in the sense that they depend on the *parameters* in a simple linear fashion (even if the relationship between X and Y is strongly nonlinear).

One of the most widely used generalized linear models is logistic regression. In logistic regression, the Y variable is a 0 or 1 Bernoulli variable (you can also think of this as a yes or no/heads or tails). In logistic regression, the goal is to predict a yes or no answer using a linear function of X. Obviously, random observations from a pool of 0s and 1s are not modeled very well by a Gaussian distribution. In order to have a model that fits well to this type of data, we would like to plug a (linear) function of the b's and X's into a function that produces a value between 0 and 1. In logistic regression, this function is the logistic function. In the simplest logistic regression, we have

$$E[Y|X] = \frac{1}{1 + e^{-(b_0 + b_1 X)}} = P(Y = 1|X)$$

Notice that I already wrote the probability of $Y = 1$ in this model directly: the expectation of a 0 or 1 variable is simply the probability of 1 (see Exercises for Chapter 6). Since there are only two possibilities, the probability of 0 is simply $P(Y = 0|X) = 1 - P(Y = 1|X)$. In this model, $E[Y|X]$ is no longer the mean of a Gaussian distribution, but rather the single parameter of a Bernoulli distribution. The likelihood is then

$$P(Y|X, b_0, b_1) = \prod_{i=1}^{n} \left[\frac{1}{1 + e^{-(b_0 + b_1 X_i)}} \right]^{Y_i} \left[1 - \frac{1}{1 + e^{-(b_0 + b_1 X_i)}} \right]^{1 - Y_i}$$

It is straightforward to take the log and set derivatives equal to 0. However, as far as I know, there are no simple closed form solutions for the MLEs. The good news is that this likelihood function can still be maximized reliably using numerical optimization methods, so it's usually no problem to fit logistic regression models using modern statistics packages.

Generalized linear models can be used to predict other distributions besides 0 and 1 variables. For example, if the "link" function is Poisson, the mean parameter of the Poisson distribution is given by a linear function of X, and the Ys can be natural numbers 0, 1, 2, ..., ∞. A more powerful generalization of logistic regression is multinomial regression where Y takes on one of several options (not just yes or no).

EXERCISES

1. What is the difference between a bivariate Gaussian model with a positive term in the off-diagonal term in the covariance matrix and a linear regression model with b_1 greater than 0?

2. Why did I derive only two MLEs for linear regression when there are three parameters in the model?

3. Draw graphical models representation of the conditional dependence structure for simple linear regression and logistic regression.

4. Write the likelihood for a regression-like model where Y is assumed to have a Poisson distribution (as opposed to the standard Gaussian). Why is this model more realistic for analysis of the relationship between mRNA levels and protein levels as in the example discussed above?

5. Write the log-likelihood for the regression model $E[Y|X] = b_0 + b_1 X + b_2 X^2$ with Gaussian distributed errors, and derive a formula for the MLE by taking the derivative with respect to b_2, setting it to 0, and solving for b_2.

6. Why is Gaussian noise in log space considered "multiplicative?" (*Hint*: Another way to write the simple regression model is $P[Y|X] = b_0 + b_1 X + N(0,\sigma)$.)

7. Show that Kernel regression converges to a simple average in the limit of infinite bandwidth and a Gaussian Kernel. What happens to the Kernel regression estimate when bandwidth goes to 0?

8. Take derivatives of the weighted SSR with respect to the parameters and set them to zero to find out how to maximize the weighted SSR in LOESS.

REFERENCES AND FURTHER READING

Cleveland WS, Devlin SJ. (1988). Locally weighted regression: An approach to regression analysis by local fitting. *J. Am. Stat. Assoc.* 83(403):596–640.

Csárdi G, Franks A, Choi DS, Airoldi EM, Drummond DA. (May 7, 2015). Accounting for experimental noise reveals that mRNA levels, amplified by post-transcriptional processes, largely determine steady-state protein levels in yeast. *PLoS Genet.* 11(5):e1005206.

CHAPTER 8

Multiple Regression

So far, we've been thinking about regressing one list of numbers on another, but this is only the very simplest use of linear regression in modern molecular biology. Regression can also help us think about the kinds of high-dimensional data vectors that we were considering in the context of clustering and sequence analysis in Chapters 5 and 6. The major way regression is used in this context is when we reinterpret X as a high-dimensional set of variables that might explain Y.

8.1 PREDICTING Y USING MULTIPLE Xs

A good motivation for multiple regression is the data we've already seen in Chapters 3 and 7 for quantitative trait loci for gene expression or eQTL analysis: We might be trying to explain gene expression level as function of genotypes at many loci that we've measured for that individual. We want to see if the level of a gene Y is associated with the genotype at a combination of loci.

Multiple regression will allow us to make predictions about Y based on multiple dimensions of X. Notice that in this example (Figure 8.1), the relationship is not strictly linear: The effect of the genotype on chr8 only has an effect on gene expression if the cells have the "a" allele on chr3. This type of behavior can be included in a linear model by including another dimension of X that is set to be the product of the genotypes at the two loci: If the genotypes are represented as 0 and 1 (e.g., for a and A), the product is only 1 when an individual has 1 genotype at both alleles. In this example, there would be a negative coefficient associated with the product of the genotypes: *SAG1* expression is lower if you have 1 at both alleles. However, even without this

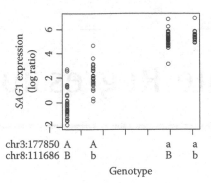

FIGURE 8.1 The expression level of the yeast *SAG1* gene is associated with the genotypes at two loci. To regress a phenotype (like gene expression) on genotype, we use 0 and 1 (and 2 for a diploid) to represent the possible genotypes at each locus. (Data from Brem, and Kruglyak, 2005.)

interaction term, it is totally reasonable to test for an association between both alleles and the gene expression level using multiple regression.

To define multiple regression more formally, we now have a vector of "features" or "covariates" (*X*s) that we are trying to use to explain a "response" (*Y*). Once again, we will assume that they are related by a simple linear model so that

$$E[Y_i|X_i] = b_0 + b_1 X_{i1} + b_2 X_{i2} + \cdots + b_m X_{im} = b_0 + \sum_{j=1}^{m} X_{ij} b_j = X_i b$$

where I have included the index *i* for the *i*th observation of *Y* (which is so far assumed to still be a simple number) and introduced *j* to index the *m*-dimensional feature vector X_i. Notice that as it is typical with multivariate statistics, I have tried to write this conveniently using linear algebra. However, in this case it's a bit inelegant because I have to add an extra 1 at the beginning of the feature vector *X* to make sure the b_0 term is included, but not multiplied by the first dimension of the *Z*. Figure 8.2 shows the structure of the multiple regression problem.

Now we can write out a Gaussian likelihood for the regression where the mean is just $X_i b$, and everything else is the same as in the univariate case:

$$L = P(Y|X, \theta) = \prod_{i=1}^{n} N(Y_i | X_i b, \sigma)$$

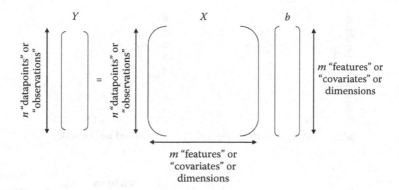

FIGURE 8.2 The multiple regression problem expressed using linear algebra notation. The response, Y, is predicted to be the product of the data matrix, X, and a vector of coefficients, b. The sizes of the vectors and matrices correspond to the number of datapoints or the number of features.

Note that all the assumptions of univariate regression still hold: We are still assuming the features, X, are measured perfectly (no noise) and that the observations, indexed by i, are i.i.d.

8.2 HYPOTHESIS TESTING IN MULTIPLE DIMENSIONS: PARTIAL CORRELATIONS

Multiple regression seems like a great way to identify which features (or covariates) are predictive of a response. In the case of genotypes and phenotypes (as in eQTL data), the interpretation is very clear: The genotypes on different chromosomes are independent, and they cause the phenotype (or gene expression level). It's not likely (given our understanding of molecular biology) that genotypes (at one or more loci) could be caused by the gene expression level, or that a mutation at one locus would cause a mutation to appear at another. Therefore in this case, the arrow of causality is pretty clear. However, as I pointed out in Chapter 7, regression can also be used to formulate powerful tests of association that are symmetric between the Y and X (e.g., to reject the null hypothesis that the correlation is 0). In many cases, we might not know which variable causes the other. This becomes particularly complicated when we have many covariates (or features, X) that are correlated with each other.

One great example of this is a long-running debate about what causes the variation in the rate of protein evolution. For example, it has been

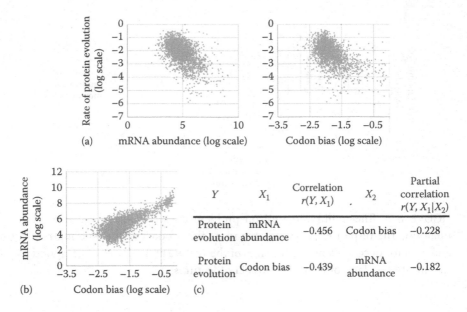

(a) mRNA abundance (log scale)

(b) Codon bias (log scale)

Y	X_1	Correlation $r(Y, X_1)$	X_2	Partial correlation $r(Y, X_1 \mid X_2)$
Protein evolution	mRNA abundance	−0.456	Codon bias	−0.228
Protein evolution	Codon bias	−0.439	mRNA abundance	−0.182

(c)

FIGURE 8.3 Partial correlations and hypothesis testing: (a) mRNA abundance (left) and codon bias (right) are both strongly associated with the rate of protein evolution. Each gray symbol represents a yeast gene. (b) mRNA abundance and codon bias are also strongly correlated. Thus it is unclear if they are independently associated with the rate of protein evolution. (c) After correcting for the other covariate using partial correlation analysis, we can see that both mRNA abundance and codon bias appear associated with the rate of protein evolution, albeit much more weakly that the standard correlation suggests.

repeatedly observed that highly expressed genes have slower rates of evolution (Figure 8.3a). Highly expressed genes might be more likely to impact cellular function, and therefore might be more highly conserved over evolution. However, maybe protein evolution is biased by a difference in the mutational properties of these proteins—indeed, the genes that evolve slowly also tend to have the strongest codon bias. So maybe proteins evolve more slowly because there are fewer codons that are acceptable, or because they have more chances at transcription coupled repair. To make matters more confusing, genome-wide mRNA expression levels are strongly associated with the codon bias in genes (Coghlan and Wolfe 2000). This result is reproduced in Figure 8.3b, and the relationship is undeniable (data we collected by Drummond et al. 2006 and Wall et al. 2005). So how do we figure out what is really correlated with what?

Luckily, there is a very simple way to correct for the effect of a third variable on the correlation between two other variables, known as the "partial" correlation (in analogy with the "partial" derivatives that we have been using on multivariate functions in all of our derivations). For example, if you want to know the correlation between Y and X_1, while holding X_2 constant, you can first do a regression of Y on X_2, then do a regression of X_1 on X_2. If there is a correlation between the residuals of these regressions, then you know there is a relationship between Y and X_1, independent of X_2. You can calculate the significance of the correlation of residuals just as you would to test for association between any two variables (as we saw in Chapter 7) although you are now assuming that the joint distribution of all three variables is Gaussian. It turns out there is a shortcut formula to calculate the partial correlation, so you don't actually have to do the two regressions.

$$\rho(Y, X_1 | X_2) = \frac{r(Y, X_1) - r(Y, X_2)r(X_1, X_2)}{\sqrt{1 - r(Y, X_2)^2}\sqrt{1 - r(X_1, X_2)^2}}$$

where
  r is the standard Pearson correlation coefficient between two variables
  ρ is the partial correlation between two variables *given* a third variable

This formula was probably more useful in the days before statistics software—it's now almost as easy to just calculate the correlation between the residuals of two regressions.

The interpretation of the partial correlation is that it summarizes the correlation between X_1 and Y after the relationship between each of them and X_2 has already been taken into account. In our case, the partial correlations we get are summarized in the Figure 8.3c. It turns out that they are both highly significant: Although the relationship between mRNA levels and codon bias is very strong, it does not appear to fully explain the correlations we see between these variables and protein evolution. The partial correlation between protein evolution and each of these variables is 0.2, which is still strongly significant in both cases ($t < -9$, $P < 10^{-10}$).

Another way we can try to test for independent effects of the two variables is to make a multiple regression model, where we write

$$E[Y_i|X_i] = b_0 + b_1 X_{i1} + b_2 X_{i2}$$

where
  Y_i is the log of the rate of protein evolution
  X_{i1} is the log of the mRNA level
  X_{i2} is the log of the codon bias for gene i

As I mentioned before, it's possible to dispense with the assumptions of the probabilistic interpretation of linear regression by using the least squares interpretation of the objective function. In this case, the multiple regression model including both mRNA levels and codon bias has $R^2 =$ 0.235 (explains 23.5% of the variance in the rate of protein evolution). The mRNA levels alone explain 20.8%, while the codon bias alone explains 19.3%. So naïvely, it seems like both factors are contributing something because the multiple regression model explains more than the two univariate models.

However, it's important to note that as we add more dimensions of X, the variance explained will always increase: The model with more parameters will always be able to fit the data a little bit better, even if the additional variables are capturing only a little bit of noise. We have to assess whether the improvement in variance explained is worth the additional parameter. We'll later see how to do this using the Akaike information criterion (AIC). It turns out that in the case of protein evolution, mRNA abundance and codon bias, the story is probably more complicated. It is currently believed that even more sophisticated regression techniques than partial correlation and multiple regression are needed in order to get to the bottom of these correlations (Drummond et al. 2006).

8.3 EXAMPLE OF A HIGH-DIMENSIONAL MULTIPLE REGRESSION: REGRESSING GENE EXPRESSION LEVELS ON TRANSCRIPTION FACTOR BINDING SITES

To illustrate an elegant application of multiple regression in genomics, let's try to build a statistical model to predict genome-wide expression levels from DNA sequences. A key idea in molecular biology is that gene expression levels are due in part to the presence of consensus binding sites for sequence-specific DNA binding proteins in the noncoding DNA adjacent to genes. But how much of genome-wide expression patterns do these binding sites really explain?

We can use multiple regression to try to answer this question starting with the simplest possible assumptions. Let's let the gene expression level for a gene, g, be Y_g in a given condition. If we know the binding specificities (or motifs) of transcription factors (e.g., from in vitro binding experiments), we can count up the number of times each of these sequence patterns occurs nearby each gene. In the regression model, the number of matches to each binding motif nearby the gene is the Xs that we will use to try to explain Y, the expression level. The regression model is

$$E[Y_g|X_g] = b_0 + b_{TBP}X_{g,TBP} + b_{Pho4}X_{g,Pho4} + b_{Msn2}X_{g,Msn2} + \cdots$$

where for each transcription factor we have a parameter b, and X are the numbers of matches to the binding motif for each transcription factor (TBP, Pho4, Msn2, etc). Notice that as with the eQTLs, the Xs in this model are not well-modeled by Gaussian distributions. The number of matches nearby each gene can be 0, 1, 2, ... Again, we won't worry too much about this because the regression model assumes that there is Gaussian noise only in the Y, and that the Xs are measured perfectly. We'll use the log ratios of gene expression changes in this model, which are real numbers (like −0.24, 1.4345, etc.), so we expect the Gaussian noise assumption to work reasonably well.

It is interesting to think about the biological interpretation of the vector, b, in this model. If the component of b corresponding to Msn2 is 0, the presence of that binding site does not predict anything about expression levels in the condition in which the experiment was conducted, perhaps suggesting that the transcription factor is inactive. On the other hand, a positive b means that genes with that sequence pattern nearby tend to be relatively overexpressed, perhaps suggesting that the transcription factor tends to activate gene expression. Similarly, a negative b might indicate that a transcription factor is a repressor, leading to lower levels of gene expression. Since the model is *linear*, we are assuming that all transcription factors that happen to bind near a gene combine independently to determine its gene expression, and that the impact that a transcription factor has on that gene's expression is proportional to the number of binding sites for that transcription factor nearby. Of course, these assumptions are very naïve, but the point here is that if we make these assumptions it becomes straightforward to actually test the model *globally and quantitatively.*

If we go ahead and do this multiple regression, we can compute the R^2, which tells us how much of the total variation in gene expression measurements can be explained under the simple additive assumption.

For example, if we try regressing gene expression changes due to growth in low phosphate conditions on the transcription factors mentioned earlier, we find that all three have b's that are significantly greater than 0 (activate transcription?), and that together they can explain about 3.7% of the total variation in gene expression due to low phosphate conditions (Figure 8.4). Now, 3.7% might not sound like a lot, but when interpreting this we have to consider a few things: First of all, how much of the variation could *possibly* have been explained? The data I used were from early microarray experiments (Ogawa et al. 2000) that undoubtedly contain a lot of noise: The R^2 between biological replicates is only about 13%, so the model might account for more than 1/4 of the variance that could have been explained. Whenever you are building a statistical model, it's always good to keep in mind that the explainable variance might be much less than the total variance in the data: If your model starts fitting the data better than the agreement between replicates, this probably means that you're overfitting. The second thing that it's important to remember is that we only used three transcription factors to explain *all* the gene expression changes. In fact, the cell has hundreds of transcription factors and many of them are active at any given time.

It's well known that transcription factors can work together to encode so-called "combinatorial" logic. For example, it might be important to have both the TBP and Pho4 binding sites in the noncoding DNA in order

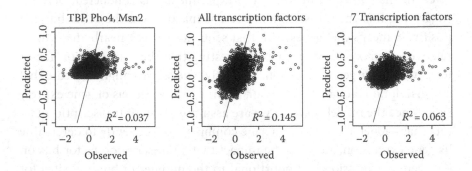

FIGURE 8.4 Predicted and observed gene expression data. Predictions or fitted values of the regression model, $E[Y|X]$ (vertical axis) compared to the observed values, Y (horizontal average), for three multivariate regression models.

to achieve full repression of gene expression. In regression, this type of logic can be included as an "interaction" between two covariates (or features) by defining a new covariate that is the product of the two covariates. For example, in this case the regression might be

$$E[Y_g|X_g] = b_0 + b_{\text{TBP}}X_{g,\text{TBP}} + b_{\text{Pho4}}X_{g,\text{Pho4}} + b_{interaction}X_{g,\text{TBP}}X_{g,\text{Pho4}}$$

where I have included the coefficient $b_{interaction}$ for the interaction term. Conveniently, we can test whether there is statistical evidence for the interaction by testing if $b_{interaction}$ is different from zero. It's important to notice that including these kind of interactions increases the number of parameters in the model, so that measures like R^2 will always increase as more such interactions are included in the model. Indeed, including interaction terms in the phosphate gene expression model does increase the R^2 by about 1%, but the interaction terms are not significant. With a large number of covariates the number of possible interactions becomes very large, and trying to include them all creates a very bad multiple-testing problem. Finally, you should notice that including interactions in this way assumes that the interaction is proportional to the product of the covariates. In practice, this might not be true.

IDENTIFYING TRANSCRIPTION FACTOR BINDING MOTIFS USING THE REDUCE ALGORITHM

Interestingly, the original application of this model (now a classic bioinformatics paper) was to consider the case where the consensus transcription factor binding sites are unknown. In this case, it's possible to try to do the regression using all possible short DNA sequence patterns (say all DNA 5-mers, 6-mers, and 7-mers) as "potential" consensus binding patterns, and test them all. Because the linear regression model has analytic solutions, it is no problem to do these calculations on a computer. However, if you test enough possible DNA sequence patterns, eventually you will start to find some that can explain some of the gene expression levels by chance. In this case, it's very important to think about multiple-testing corrections to decide which of the DNA sequence patterns are correlated with the expression data more than you would expect by chance.

In the original REDUCE paper (Bussemaker et al. 2001), the authors applied the linear regression model to many gene expression experiments and obtained measures of statistical associations for each short DNA sequence pattern for each experiment. They then had to choose a few of the most significant patterns to include as they built up the model. They used an iterative approach where they asked if each new motif explained more additional variance than would be expected for a random motif.

What if we didn't know which transcription factors were important for the response to low phosphate conditions in this example ? We could try to include all of the transcription factors in a database in our model. For example, here are the p-values that I got from regressing gene expression on motif matches for ~200 transcription factor motifs from the YETFASCO database (De Boer and Hughes 2012, Figure 8.5).

Based on this distribution, I would guess that there are about 20 or 25 transcription factors that are contributing significantly to the model. At first, it seems reasonable that about this many transcription factors would be involved in the genome-wide expression changes in response to low phosphate. However, one important consideration is that not all of the transcription factor motifs are independent. For example, in the 3-transcription factor model, Msn2 had $b = 0.056$, meaning that for every match to the Msn2 motif, the expression of that gene was about 5.7% higher. However, if we now include Gis1 to make a 4 transcription factor model, although all 4 transcription factors are still highly significant, we find that the parameter for Msn2 is reduced to $b = 0.035$. It turns out that the binding motifs for these 2 transcription factors are very similar: When a DNA sequence has a match for Gis1 it very often also has a match for Msn2. Indeed, the number of matches for Gis1 and

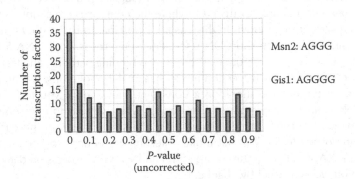

FIGURE 8.5   Distribution of P-values for the multiple regression of gene expression levels on motif matches ~200 transcription factors. This model explains about 14.5% of the variance of the gene expression data, which is probably more than the total amount of explainable variance in the data: The model is overfit.

Msn2 are highly correlated ($R^2 = 0.26$). The regression model can use either the Gis1 matches or the Msn2 binding matches or both to explain the expression levels.

As we have seen in Figure 8.3, correlations between features (or covariates) are typical in biology, especially when the number of features is large. In cases where the number of features is large, including multiple correlated features can usually improve the fit to the model slightly because the features are not usually *perfectly* correlated, so the model can always fit a little bit more of the noise, or use the two variables to compensate for the data deviations from the linear assumption. In principle, you could try to test all of the correlated features using partial correlation analysis (described earlier). However, because it works on pairs of variables, it's not really practical to apply it to a large number of covariates. In general, when building a high-dimensional model, one has to be cautious about including correlated features, because the tendency of regression models is to add more parameters and improve the fit.

8.4 AIC AND FEATURE SELECTION AND OVERFITTING IN MULTIPLE REGRESSION

As the number of features (or covariates) becomes larger than a handful, one often wants to test which (if any) of the features are significantly associated with the response (Y). In practice there are several solutions to this so-called "feature selection" problem in multiple regression. The major problem in feature selection is that each new additional feature will always improve the fit of the model because the additional parameters can always explain something (even if it is really only noise). This leads to overfitting, where a model can fit the data it was trained on very well, but have little predictive power on new data.

To illustrate the problem of overfitting, let's consider fitting multiple regression models to predict gene expression from transcription factor binding sites again. However, this time, we'll use different experiments were we have biological replicates. We'll estimate the parameters of the model on one replicate (training data) and then we'll measure the fit (the fraction of variance explained) on both the training data, and the other replicate (test data). Because regression models are fast to compute, we'll go ahead and build models starting with the first transcription factor in the yeast genome, adding one more with each model until we get to the 216th, which corresponds to the model given earlier with all the transcription factors.

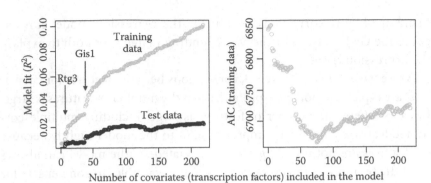

FIGURE 8.6 Overfitting in multiple regression and the AIC. The left panel shows the fit of the multiple regression model as a function of the number of dimensions of X used for the prediction. The gray points represent the fit on data used to estimate the parameters of the model, while the black points represent the fit on data from a replicate experiment. The right panel shows the AIC (on the data used to fit the model) as a function of the number of dimensions of X.

As more dimensions of X are used to predict Y, you can see that the fit to the training data always improves (Figure 8.6). The fit to the test data (that was not used in estimating the parameters of the model) plateaus and may even decrease. This is because the additional parameters are fitting noise in the training data that is not reflected in the held-out data. Two clear increases in fit occur for both training and test data when Rtg3 (a transcription factor with a similar motif to Pho4) and Gis1 (a transcription factor with similar motif to Msn2) are added to the model. Because the increase in model fit occurs for both the training and test data, we infer that these are real predictors of the gene expression response. You can see that the model might predict as little as about 2% of the biological variation, even though it appears to predict 10% on the training set. I say that it *might* only predict that little because that assumes that the biological replicates really are replicates and any nonrepeatable variation is noise. There could easily be unknown factors that make the two experiments quite different.

Although I have presented the problem of overfitting in the context of multiple regression, it is a theme that applies (at least in some form) to every statistical model (or algorithm) that estimates (or learns) parameters from data. As we learn a more and more complex model, we can improve prediction of the training set. However, the prediction accuracy of our model on new data (sometimes referred to as generalizability) of our model may not increase, and might even decrease.

The classical statistical approach to avoid overfitting in linear regression is to add the additional covariates sequentially. As each new dimension of X is added to the model, all the previous ones are tested to see if the new feature captures their contribution to the model better. If so, previous Xs are removed in favor of the new one. Only if the new variable makes an *additional* statistically significant improvement to the fit is it retained in the final model. A major limitation of this approach is that the order that the features are added to the model can affect the result, and it's tricky to perform multiple testing corrections in this framework: The number of tests must be related to the total number of features tried, not just the ones actually included in the model. Rather than statistically testing the covariates one by one, we'd prefer to measure the improvement of fit for each additional covariate and decide if the improvement in fit is "worth" the extra parameters. Because the objective function in regression is a likelihood, we could try to use the AIC (described in Chapter 5) to choose the best model, taking into account the number of parameters.

To illustrate the power of the AIC for combating overfitting, I plotted the AIC for each of the transcription factor models here. To be clear, the AIC is computed based on the maximum likelihood, which is obtained during the parameter estimation. Notice that unlike the fit to the training data, the AIC does not continue to increase as we add more and more parameters to the model. In fact, it reaches its minimum with around 75 transcription factors, and encouragingly, this is where the fit of the model to the (unseen) test data reaches its plateau. So of the 216 models that we tried, the AIC does seem to be telling which model captured the most real information about the data, without including additional dimensions that were irrelevant for prediction of new data. An important subtlety about the minimum AIC model is also illustrated very clearly: Just because the model has the minimum AIC, and therefore is in some sense the "best" trade-off between data fit and model complexity, this does not imply that the fit (or predictive power or accuracy) that we measured on the training data actually reflects the fit of the model to new data (generalization accuracy). Indeed, in this case, the fit is still badly overestimated.

The AIC is a very powerful and general way to compare models with different numbers of parameters. In practice, however, this approach only works when there are relatively few Xs (or features) to try. In the case of hundreds of possible features, it's simply impractical to try all possible combinations of features and choose the one with the smallest AIC. In the previous example, I arbitrarily chose 216 models adding transcription factors one at a time starting with chromosome 1. We have absolutely no guarantee that the 75 transcription factor model that minimizes the AIC is the only model that achieves that AIC: The 7 transcription factor model I showed based on transcription factors I thought might be important has nearly the same AIC as the 75 parameter model that happened to be the best of the 216. Furthermore, there are likely to be models with much lower AICs: To be sure that we found the lowest, we would have to try all possible combinations of the 216, an astronomically large number. Even though multiple regressions are very fast to calculate, trying 216! combinations (or even all possible 3 transcription factor models) would be computationally prohibitive. Not to mention the multiple testing problem. Finding the best set of features in a model is known as "feature selection" and is a fundamentally difficult problem for high-dimensional models.

In general, feature selection is the analog of multiple hypothesis testing when we are building multivariate predictive models. Unfortunately, there is no solution as comprehensive and effective for feature selection as FDR and Bonferroni corrections are for multiple hypothesis testing.

MULTIPLE REGRESSION SO FAR

- Linear regression has a straightforward generalization if the data used to predict is more than one-dimensional.
- Partial correlations can be used to distinguish effects for two correlated covariates.
- Overfitting is when a model with lots of parameters has been trained on one dataset and predicts that data very well, but can't predict well on data that wasn't used for training. This usually indicates that the model has learned to predict the particular noise in the training data. This problem gets worse as the number of parameters that need to be estimated grows.
- The AIC is a simple and effective way to choose between models with different numbers of parameters, as long as the number of models doesn't get too large.

EXERCISES

1. What is the biological interpretation of the parameter b_0 in the gene expression transcription factor binding site regression model as formulated here?

2. Assume that you wanted to test all DNA patterns of length $w = 6$ for their association with gene expression. Explain how you could set up the FDR correction for this problem?

3. How many parameters are estimated in the linear regression model with 216 transcription factors?

4. It took 2 minutes for my laptop to fit (using R) all 216 multiple regression models I showed in the figures. Assuming each model takes about the same amount of time to fit, how long would it take to find the best possible model in this example?

5. PBMs (protein binding microarrays) measure the binding affinity of DNA-binding proteins to every possible DNA sequence (up to length 10). Formulate the prediction of DNA-binding affinity as a multiple regression problem (what is Y? what is X?).

REFERENCES AND FURTHER READING

Brem RB, Kruglyak L. (2005). The landscape of genetic complexity across 5,700 gene expression traits in yeast. *Proc. Natl. Acad. Sci. USA* 102(5):1572–1577.

Bussemaker HJ, Li H, Siggia ED. (2001). Regulatory element detection using correlation with expression. *Nat. Genet.* 27(2):167–171.

Coghlan A, Wolfe KH. (2000). Relationship of codon bias to mRNA concentration and protein length in *Saccharomyces cerevisiae*. *Yeast* (Chichester, England) 16(12):1131–1145.

De Boer CG, Hughes TR. (2012). YeTFaSCo: A database of evaluated yeast transcription factor sequence specificities. *Nucleic Acids Res.* 40(Database issue):D169–D179.

Drummond DA, Raval A, Wilke CO. (2006). A single determinant dominates the rate of yeast protein evolution. *Mol. Biol. Evol.* 23(2):327–337.

Ogawa N, DeRisi J, Brown PO. (2000). New components of a system for phosphate accumulation and polyphosphate metabolism in *Saccharomyces cerevisiae* revealed by genomic expression analysis. *Mol. Biol. Cell* 11(12):4309–4321.

Wall DP, Hirsh AE, Fraser HB, Kumm J, Giaever G, Eisen MB, Feldman MW. (2005). Functional genomic analysis of the rates of protein evolution. *Proc. Natl. Acad. Sci. USA* 102(15):5483–5488.

CHAPTER 9

Regularization in Multiple Regression and Beyond

I N THE DISCUSSION OF multiple regression in Chapter 8, we saw that as the number of dimensions X becomes large, we have to ensure that additional parameters in our model are not overfitting the data. The classical statistical solution to this problem is to add covariates (or features, or dimensions of X) sequentially and test whether the new parameter estimated is statistically significant. If so, then test all the previous covariates and see if any of their parameters are no longer significant with the new covariate included. Needless to say, this becomes tedious if there are more than a handful of features to be considered.

A somewhat more practical solution that I presented in Chapter 8 is to fit several models with different numbers of parameters and compare them using the Akaike information criterion (AIC). Although this can be applied more generally, we still need to fit and then compute the AIC for all possible models. As we saw in Chapter 8, however, this is still not very feasible for hundreds of features (or more) that we typically have in genome-scale quantitative biology.

The use of the AIC to trade-off model complexity (number of parameters) against likelihood suggests that there might be a smarter way to fit complex models. Rather than maximizing the likelihood of many different models and then computing their AICs, we might try to maximize a different objective function that has the property we like about the AIC, namely, that it tends to choose the models with fewer parameters unless the difference in likelihood between the models is large enough.

Another way to think about this problem is to consider that when we fit the multiple regression model, we are not using a key piece of information: We expect most of the dimensions of X to be unrelated to our prediction problem. Said another way, we expect most of the components of the "b" vector to be 0. In machine learning jargon, we expect the parameter vector to be "sparse," meaning most of the values are 0s. Ensuring sparsity ensures a small AIC because it means that we don't include most of the possible parameters in the model. Of course, the problem is that beforehand, we don't know *which* of the dimensions should have their parameters set to 0, and which ones we should include in our model. And as I pointed out in Chapter 8, the number of possible combinations means that for high-dimensional problems, it's infeasible to try them all.

9.1 REGULARIZATION AND PENALIZED LIKELIHOOD

This brings us to one of the most powerful ideas in machine learning: regularization. Regularization can solve the problem of overfitting by allowing us to fit simpler models based on very high-dimensional data. The key idea is to use regularization to set most of the parameters equal to zero. Here, we'll consider one very elegant type of regularization for probabilistic models, called "penalized likelihood," where we modify the objective function by adding new terms to the likelihood. In doing so, we hope to encourage the model to set most of the parameters to zero. In other words, penalized likelihood methods amount to changing the objective function, in this case so that the objective function deems models to fit the data better if they have few nonzero parameters. It's important to remember that regularization is a more general term, and even models that don't have probabilistic interpretations can have regularization schemes to encourage sparsity and avoid estimating too many parameters from too little data.

In the case of linear regression, we typically consider penalized likelihoods of the following form:

$$PL = \log[P(data|\theta)] - f(\theta) = \log[P(Y|X,\theta)] - \lambda_1 \sum_{j=1}^{m} |b_j| - \lambda_2 \sum_{j=1}^{m} b_j^2$$

where the first formula is the general formula for a penalized likelihood (PL), where I have written the penalty as an arbitrary function, f, of the parameters, theta. In the second formula, I showed the kind of penalized likelihood that's typically used as an objective function for linear regression. You can see that in the penalty terms for linear regression subtract from the likelihood based on the absolute value of slope parameters of the linear regression model. Note that when b parameters are 0, there is no penalty. If a b parameter in some dimension is not zero, the penalty is weighted by the parameter lamda, which controls the relative strength of the penalty. In the machine learning jargon, the first penalty term is referred to as an $L1$ regularization because it is proportional to the $L1$ norm of the vector b. The second penalty term is an $L2$ regularization because it is proportional to the $L2$ norm of the vector. In practice, a mixture of these two (the so-called "elastic net" Zou and Hastie 2005) or $L1$ alone (the "LASSO" Tibshirani 1996) work well for linear regression.

Terminology aside, I hope that the form of this new objective function makes some sense: The penalized likelihood is explicitly trading off the fit to the data (the likelihood) against a penalty function based on the parameters. The more the parameters deviate from 0, the more gets subtracted from the likelihood. The larger the λ parameters, the bigger the change in likelihood needs to be before a nonzero parameter can be accepted into the model. In the limit of very small λ, the penalized likelihood objective function reduces to the ordinary likelihood.

It's also worth noting that adding penalty terms to the likelihood (in particular the $L1$ penalty) means that there are no longer analytic solutions for the b parameters; the models must be fit using numerical methods. This means that to use them in practice, you'll always be relying on the efficiency and accuracy of software packages. Be sure to check for the latest and greatest implementations. For example, there are several packages implementing these models in R. In general, penalized likelihoods can lead to computationally hard optimization problems; this means that it is often not possible to optimize these objective functions in a practical amount of time for very large datasets with millions of datapoints. However, for the specific case of regression (including generalized linear models), it's known that the maximum of the penalized likelihood objective function can be identified efficiently (Friedman et al. 2010).

Fitting models using penalized likelihoods can in principle solve the problem of model complexity/feature selection for linear regression. Estimating parameters using $L1$ regularized likelihood tends to produce models with much fewer nonzero parameters than standard linear regression, and the parameters that are nonzero tend to be much more strongly associated with the data. Another great advantage of penalized regression is that it's possible to fit models where the number of covariates (predictors, dimensions of X) is larger than the number of observations. This is in contrast to standard multiple regression, where if the number of dimensions is greater than the number of observations, the estimation procedures will fail. However, one inconvenience of regularized regression compared to standard regression is that because the parameter estimates are no longer maximum likelihood estimates (they are maximum *penalized* likelihood estimates) we can no longer apply well-developed theory that we used to derive the distribution of parameter estimates and P-values for hypothesis tests of association that we saw in Chapter 6. Of course, distributions of parameters and P-values for hypothesis tests can still be obtained numerically using permutations and resampling (see Chapter 2).

Perhaps the more serious drawback of the penalized likelihood objective functions that I've described is that although they can ensure that most of the b parameters are held at 0, to do so, I had to introduce two additional parameters: λ_1 and λ_2. These parameters also need to be estimated—I hope it's obvious that if we tried to choose them to maximize the objective function, we would simply set them to 0. The penalized likelihood is always less than the likelihood. In practice, therefore, these parameters are usually chosen using a cross-validation procedure. The idea of the cross-validation procedure is to estimate the parameters of the model on a subset of the data, and see how well it predicts the data that was left out during the estimation. (We will return to cross-validation later in the book in our discussion of evaluation of classifiers.) Cross-validation allows the regularization parameters to be chosen so that the model is producing the best possible predictions of data that wasn't used in the training, and therefore can't be "overfit." Of course, the downside of this cross-validation procedure is that we need enough data to actually leave some of it out during training. With thousands of datapoints, this isn't usually a problem for continuous regression problems, but it can become an issue for logistic and multinomial regression in the classification setting. We will return to this in the next chapters. If there isn't enough data for cross-validation, it's possible to choose a regularization parameter that produces a number of nonzero parameters that matches some biological knowledge or intuition. We'll return to the intuition behind these regularization parameters later in this chapter.

Applying regularization to the transcription factor regression problem (discussed in Chapter 8) allows us to automatically obtain sparse models in an unbiased way. For example, with an $L1$ penalty of 0.045, I got a 3-transcription factor model with R^2 of about 0.029 on the training data, and 0.011 on the replicate. The model has an AIC of 6722, which compares pretty well to what I got with the models I trained with a lot more work in Chapter 8.

9.2 DIFFERENCES BETWEEN THE EFFECTS OF $L1$ AND $L2$ PENALTIES ON CORRELATED FEATURES

Although the form of the penalties I proposed is somewhat arbitrary at this point, it's useful to consider the effect that the two penalties would have on highly correlated features. For example, let's imagine the extreme case where two features, X_1 and X_2 are *perfectly* correlated, and the b for either of them alone is 0.5. In this case, either one could be used to predict Y, and there's no additional information gained by including the other one. If we use the $L2$ penalty, the optimal solution is to assign b = (0.25, 0.25), which gives a total penalty of 0.125λ. Any unequal distribution to the two features gives a larger penalty. This means that the $L2$ penalty prefers to share the predictive power amongst the correlated features. In practice, this means that the $L2$ penalty doesn't reliably produce sparse models when features are strongly correlated; instead, it just assigns smaller bs to all the correlated features.

If we use the $L1$ penalty, on the other hand, any solution where the two components add up to 0.5, such as b = (0.5, 0) or b = (0, 0.5) yield the same total penalty to the likelihood of 0.5λ. This means that the $L1$ penalty has no preference between sharing and just choosing one of the features if they are perfectly correlated. For any real features that are not *perfectly* correlated, one of the features will explain the data slightly better (possibly due to noise) and that feature will be selected by the model, while the other one will be removed from the model. Thus, it's the $L1$ penalty that's actually the one pushing the parameters to 0 when there are correlated features. In practice, however, the $L1$ penalty can be overly "aggressive" at removing predictive features, especially in molecular biology where we may wish to preserve bona fide biological redundancy.

Because of the different behavior of the two penalties on correlated features, it's thought that penalties composed of a combination of the $L1$ and $L2$ can work better than either alone. The $L1$ part works well to get rid of features that are only fitting noise, while the $L2$ part encourages inclusion of multiple highly correlated features in the model. Thus, current thinking is that a combination of both penalties is best, although this means that two regularization parameters must be chosen for each model.

9.3 REGULARIZATION BEYOND SPARSITY: ENCOURAGING YOUR OWN MODEL STRUCTURE

Although ensuring sparse models is the most common use of penalized likelihood, recent advances in machine learning have generalized the idea to develop objective functions that bias the model parameters in nearly any predetermined direction.

Let's say you were building a multiple regression model for gene expression data based on the presence of transcription factor motifs, just as we were doing in Chapter 8. However, the data available was for gene expression measurements for a time series, where expression levels were measured each hour for 5 hours. For the expression of gene, g at time, t, you might write

$$E[Y_{gt}|X_g] = X_g b_t$$

where

Y represent the expression levels for gene g at time t

X represent the presence of transcription factor binding motifs in the promoter of gene g (maybe 100s of these)

b are the regression parameters that represent inferred effects of each transcription factor on gene expression, where we assume that these can change over time—perhaps some transcription factors are active at the early time points, while others are active only at the end of the experiment

Based on the discussion in this chapter so far, and the problems with overfitting of these types of models we saw in Chapter 8, I hope you're convinced that it would be a good idea to use (at least) an $L1$ penalty, so the objective function might have the form:

$$\log[P(Y|X,\theta)] - \lambda_1 \sum_{t=1}^{5} \sum_{j=1}^{m} |b_{tj}|$$

where m indexes the number of covariates (in this case, the different transcription factor motifs that could be included in the model), and I have simply added up the penalties for each of the 5 time points. This penalty is expected to encourage the model set $b = 0$ most of the time. Only when there is a large change in the likelihood can that b take a nonzero value.

Here's where we can apply regularization to encourage another aspect of the model structure: We know that the data comes from a time series, so we don't expect the bs to be totally independent at each time point. In fact, the closer two experiments are in the time series, the closer we expect their bs to be. We can use this intuition to add a new regularization term to bias the model to choose bs that are more similar at similar time points. For example, we might add another term to the objective function so we end up with

$$\log[P(Y|X,\theta)] - \lambda_1 \sum_{t=1}^{5} \sum_{j=1}^{m} |b_{tj}| - \varphi \sum_{s=1}^{5} \sum_{t=s+1}^{5} \sum_{j=1}^{m} \frac{|b_{tj} - b_{sj}|}{t - s}$$

where s and t now both index the 5 time points in the experiment. If s and t are subsequent time points ($t - s = 1$), a penalty proportional to their difference ($\varphi|b_t - b_s|$) is subtracted from the likelihood. However, if s and t are the first and last time points in the experiment, the penalty for bs being different is only ¼ of this ($\varphi(|b_{tj} - b_{sj}|/4)$). This means that it is much easier for the model to include different bs for the same transcription factor in time points that are separated than for time points that are subsequent. This penalty is expected to encourage the model to keep b constant for each transcription factor during subsequent time points (and therefore over the entire time course). Only when there is sufficient evidence (change in the likelihood) that a b at one time point is really different from another will the model allow bs to be different at different time points. In this case, we are using regularization to encourage the model to treat the different time points largely as replicates (penalizing bs that change over time), but still allowing the bs to vary over time if the data strongly supports it.

Thus, regularization can be used to develop models of a wide variety of structures that reflect our intuition about how the experiment was performed and what sorts of patterns we are expecting. The types of structures that can be encouraged by regularization seems to currently be limited only by the creativity of the researchers. For example, recent studies have used regularization to model population structure in genome-wide association study (GWAS) analysis (Puniyani et al. 2010) and to model the cell lineage relationships in developmental gene expression data

(Jojic et al. 2013). It's important to note that optimizing these nonstandard regularized objective functions is still a nontrivial computational problem. Because of this, so far, these models have largely been used by expert machine learning research groups.

9.4 PENALIZED LIKELIHOOD AS MAXIMUM A POSTERIORI (MAP) ESTIMATION

A more elegant way to motivate the use of penalized likelihood is to develop the new objective function using a probabilistic model based on our intuition about the problem. To encourage sparsity, we would like to ensure that most of the parameters in our model are 0, and that only a few of them are allowed to be nonzero if the data strongly supports the need for the parameter. This prior knowledge about the problem can be formalized through the use of a prior distribution on b. In other words, we have a model for the components of b: Most of them should be zero, and a few can be far from zero—of course we don't know in advance which components are the ones that should be far from 0. We can therefore think of the bs as random observations from a pool, where most of the numbers in the pool are 0. There are many models that can give us numbers of this type, but a simple one is the exponential distribution:

$$P(b|\lambda) = \prod_{j=1}^{m} \lambda e^{-\lambda|b_j|}$$

where I've used the absolute values of the components, because the exponential distribution is only defined for observations greater than zero. Just to be clear what we're saying with this distribution: The *parameters* of the model, b, are going to be assumed to be drawn from a distribution where each component is independently drawn from an exponential distribution with parameter λ. A distribution on the model parameters is a prior distribution because it summarizes what we know about the parameters "prior" to seeing any data.

The simplest way to include a prior distribution in the objective function is to use the MAP objective function that I introduced in Chapter 4.

$$P(\theta|X) = \frac{P(\theta)}{P(X)} P(X|\theta) = \frac{P(\theta)}{P(X)} L$$

where I have used Bayes' theorem to rewrite the objective function. In this formula, you can see the MAP objective function has the form of a ratio times the likelihood. Since in practice we will always optimize the log of the objective function, these terms break apart and we have

$$\log P(\theta|X) = \log P(\theta) + \log L - \log P(X)$$

Substituting our exponential prior for $P(\theta)$, we have

$$\log P(\theta|X) = m \log \lambda - \lambda \sum_{j=1}^{m} |b_j| + \log L - \log P(X)$$

The first and last terms don't depend on the parameters, so they will not affect our choice of the optimal parameters (their derivatives with respect to the parameters will always be 0). Thus, up to constant terms that do not depend on b, this MAP objective function is the $L1$ penalized likelihood.

This formulation gives us a very intuitive understanding of what the λ parameter in the penalty means: It is the parameter of the exponential prior. The larger we set λ, the sharper the prior decays, and the more we penalize b values that are far from 0. In the limit of infinite λ, no amount of data will convince us to let b be anything but 0. On the other hand, in the limit of $\lambda = 0$, we don't actually prefer b values near 0. In the jargon, this last case corresponds to the case of "uninformative" or "uniform" priors, which reduces the objective function to be the likelihood. Since the λ parameters are a model for the distribution of parameters (the bs) and not data, they are referred to as "hyperparameters."

This interpretation of the λ hyperparameters as prior beliefs about the b parameters in the model suggests that we might not want to estimate them using the cross-validation procedure described earlier, but rather set them based on what we think the distribution of bs ought to be.

9.5 CHOOSING PRIOR DISTRIBUTIONS FOR PARAMETERS: HEAVY-TAILS IF YOU CAN

Although the idea of the prior distribution gives us an intuitive interpretation of the penalty parameter, we still made a seemingly arbitrary choice of the exponential distribution. Indeed, choosing other distributions will

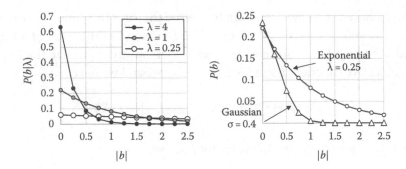

FIGURE 9.1 Illustration of priors for regularized linear regression. The left panel shows exponential prior distributions. With large prior parameter ($\lambda = 4$), there is a very strong bias to push b parameters to 0, and penalize values far from 0. However, with a small prior parameter ($\lambda = 0.25$) the penalty decays very slowly, and the prior distribution approaches uniform. Intermediate values give some tendency for the parameters to be 0. The right panel compares the shape of the Gaussian prior with the exponential prior.

imply other penalty terms. For example, a Gaussian prior leads to an $L2$ penalty (see Exercises).

In most applications, priors are chosen for mathematical convenience to try to make the penalized likelihood easy to optimize. For mathematical convenience, the so-called "conjugate" priors are always best because these ensure that the posterior distribution will have the same form as the prior distribution. However, in practice the prior distributions that ensure good final choices of parameter values are not necessarily those that are mathematically convenient. In practice, "heavy-tailed" prior distributions such as the exponential prior (or $L1$) penalty tend to work better than the very rapidly decaying priors like the Gaussian prior (or $L2$). Figure 9.1 illustrates the shapes of the prior distributions. The intuition for this is a bit subtle, but it goes something like this: Our expectation is that most of the b parameters are 0, so we want the prior distribution to be peaked at 0. However, *given* that a b is *not* zero (for the small number of nonzero parameters that we actually want to estimate) we don't expect it to be very close to 0. In fact, we want to allow the nonzero parameters to get reasonably large without adding much to the penalty. Therefore, we want the tails of the prior distribution to decay slowly, or to be "heavy". Another example of a heavy-tailed distribution is the t-distribution, which is a heavy-tailed version of the Gaussian distribution.

PRIOR DISTRIBUTIONS FOR CLUSTERING
AND INFINITE MIXTURE MODELS

In Chapter 6, I introduced the Gaussian mixture model and suggested that choosing the model that minimizes the AIC is a good way to choose the optimal number of clusters. However, if you've managed to follow this discussion of regularization and prior distributions, you are now probably wondering if there is an elegant formulation of model-based clustering that (analogous to regularization in linear regression) will choose the number of clusters automatically.

One situation where prior distributions on the number of clusters, K, have been explored is in the context of Bayesian approaches to population structure. You probably remember from my discussion of the problem in Chapter 4, that assigning individuals to subpopulations is a beautiful example where the Bayesian idea of estimating posterior distributions matches our intuition about admixed individuals, and has led to the STRUCTURE program. STRUCTURE models genotype data using a finite mixture model (one component for each subpopulation) and treats the subpopulation assignment for each individual as a hidden variable. By fitting the mixture model, STRUCTURE can infer both the population structure and the ancestry of each individual. However, a major issue with inferring population structure from genotype data is that we don't usually know the number of suppopulations for the model: We have to choose K. In the STRUCTURE manual (Hubisz et al. 2009), the authors suggest using the MAP value of K. They suggest computing the probability of the data $P(data|K)$ for several different values (say K, L, M), and then forming the ratio

$$P(K|data) = \frac{P(data|K)}{\sum_{k \in K, L, M, \ldots} P(data|k)}$$

which corresponds to a uniform prior over K. You're probably worried about this because the different models have different numbers of clusters, and therefore different numbers of parameters, so the probability of the data will always increase as K increases. And this would be true if the STRUCTURE was computing the likelihood, but, because STRUCTURE is a bona fide Bayesian method (as discussed in Chapter 4), it (hopefully) has managed to integrate out the parameters, so that the probability of the data for these different models can be fairly compared. Although this might work okay in practice for small K, it's akin to using the AIC to regularize linear regression. To use this formula, we have to actually compute the probability of the data for many different choices of K. It would be more elegant if we could formulate the model to learn the number of clusters along with the rest of the model, just as the regularized regression learned the number of b parameters that we should allow to be nonzero.

Indeed, there is just such a formulation, and it has been applied to the problem of choosing the number of hidden subpopulations in the Bayesian formulation of population structure inference (Pella and Masuda 2006, Huelsenbeck and Andolfatto 2007). However, before going further, I think it's important to consider that in linear regression, the idea of the prior distribution was to bias the parameter estimates to stay near 0. The number of parameters doesn't change; it's simply a matter of setting some of them to zero. In the case of clustering, the problem is much more complicated: If we do not prestate the number of clusters, there will be parameters in some models that do not exist in others (because there are a smaller number of clusters). Further, since we don't know the number of clusters we have to consider the possibility that there might be infinitely many: The number of clusters is a natural number (1, 2, 3, ...) Since in principle the number of clusters (and the parameters) in the model could be infinite (although this turns out to be impossible for any finite dataset), these models are sometimes called "infinite mixture models." The other name for these models is nonparametric Bayesian mixture models, which I believe is even more confusing because of the way nonparametric is used in other areas of statistics (as we used it in Chapter 2). So we will stick with the name infinite mixture models.

To understand the infinite mixture model, let's start with the likelihood for the (finite) mixture model:

$$L = P(X|\theta) = \prod_{i=1}^{n}\sum_{c=1}^{K}\pi_c P(X_i|\theta_c)$$

where c indexes the cluster number, from 1 to K, and I've separated out the parameters describing the distribution of the data (θ_c) from each cluster from the mixing parameters (π) or prior probabilities. Naively, we might try to write the MAP objective function for the mixture of Gaussians as

$$P(\theta, K|X) \propto P(K|\alpha)P(\theta|K, \nu)L$$

where $P(K|\alpha)$ and $P(\theta|K, \nu)$ are the prior distributions (or penalties) associated with the number of clusters and the other parameters. In these prior distributions, I've written additional α and ν to emphasize that these are not constant, but have shapes to encourage the structure we would like to find. In principle, we could then try to optimize both the parameters and K directly, and choose the hyperparameters by cross-validation. In practice, this turns out not to be feasible (as far as I know).

However, it turns out that if the problem is written in terms of the (unobserved) cluster assignments for each datapoint, (the Z_{ic} in our notation from Chapter 6) it is possible to derive a sampling strategy to obtain samples from the posterior distribution (Rasmussen 1999). Perhaps most important for this to work was to figure out how to choose the prior probability of an observation from one of the clusters that was already in the model or whether a new cluster should be created. If there are already a few clusters in the model, and each has a lot of datapoints assigned, we would like the prior distribution (or regularization) to tend to assign the new data to one of the clusters we already have. On the other hand, if this is one of the first data points we've observed, we'd like to give it a good chance of starting its own cluster. Thus, the prior distribution we'd like is a "rich-get-richer" process where clusters that already have a lot of data tend to get more—only if the data is very different should we allow the model to create a new cluster (and associated parameters).

This model structure can be encouraged using a so-called Dirichlet process prior. The simplest way to generate data from this process is the so-called Chinese restaurant process, where we imagine a restaurant with an infinite number of tables (the clusters), each with an infinite number of seats. As each new customer (datapoint) enters the restaurant, that person chooses a table (cluster) based on the number of other people (datapoints) already seated there or starts a new table (cluster). This process can be controlled by a single parameter, α, which is the hyperparameter that controls the growth in the number of clusters as the amount of data increases. As the number of tables in the restaurant approaches infinity, the probabilities converge to

$$P(X_i \text{ belongs to cluster } c) = P(Z_{ic} = 1 | \alpha) = \frac{\sum_{j \neq i}^{n} Z_{jc}}{n - 1 + \alpha}$$

$$P(X_i \text{ starts a new cluster}) = \frac{\alpha}{n - 1 + \alpha}$$

Note that in the top formula, $\sum_{j \neq i}^{n} Z_{jc}$ is just the number of datapoints assigned to this cluster. You can see that these rules have the desired behavior: When there are few points assigned to a cluster, X_i is almost as likely to start a new cluster as to join. On the other hand, if there are many points already in a cluster, X_i is encouraged to join. Although the mathematics of the infinite mixture is beyond the scope of this book, the sampling strategy that results is relatively simple.

EXERCISES

1. Show that in linear regression, a Gaussian (mean = 0) prior on b corresponds to $L2$ penalized likelihood.

2. What penalized likelihood corresponds to minimizing the AIC? What prior distribution does this correspond to?

3. Look up the probability density for the t-distribution. What type of penalty term would you obtain if you assume that your parameters have t-distributions?

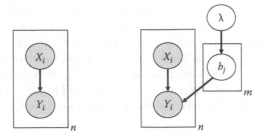

FIGURE 9.2 Graphical models representations of regression (left) and penalized regression (right). Since the b parameters are now drawn from a distribution, they are can be considered unobserved (or hidden) random variables in the model.

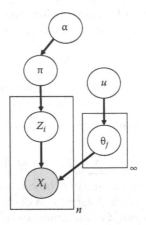

FIGURE 9.3 Graphical model representation of the infinite mixture model.

4. In the graphical model representation of penalized regression (Figure 9.2), why are the b parameters in a box with m at the bottom right?

5. Figure 9.3 shows a graphical models representation of the infinite mixture model. Look back at the graphical models representation of the standard mixture model in Chapter 6. Why are the "parameters" π and θ now explicitly represented in the diagram, whereas before they were left out?

REFERENCES AND FURTHER READING

Friedman J, Hastie T, Tibshirani R. (2010). Regularization paths for generalized linear models via coordinate descent. *J. Stat. Softw.* 33(1):1.

Hubisz MJ, Falush D, Stephens M, Pritchard JK. (2009). Inferring weak population structure with the assistance of sample group information. *Mol. Ecol. Resour.* 9(5):1322–1332.

Huelsenbeck JP, Andolfatto P. (2007). Inference of population structure under a Dirichlet process model. *Genetics* 175(4):1787–1802.

Jojic V, Shay T, Sylvia K, Zuk O, Sun X, Kang J, Regev A, et al. (2013). Identification of transcriptional regulators in the mouse immune system. *Nat. Immunol.* 14(6):633–643.

Pella J, Masuda M. (2006). The gibbs and split merge sampler for population mixture analysis from genetic data with incomplete baselines. *Can. J. Fish. Aquat. Sci.* 63(3):576–596.

Puniyani K, Kim S, Xing EP. (June 15, 2010) Multi-population GWA mapping via multi-task regularized regression. *Bioinformatics* 26(12):i208–i216.

Rasmussen CE. (1999). The infinite Gaussian mixture model. *NIPS* 12:554–560.

Tibshirani R. (1996). Regression shrinkage and selection via the Lasso. *J. R. Stat. Soc. Ser. B.* 58(1):267–288.

Zou H, Hastie T. (2005). Regularization and variable selection via the elastic net. *J. Royal Stat. Soc. Ser. B* 67:301–320.

IV

Classification

In the era of personal genomes, one of the most pressing data analysis problems is to determine which (if any) of the thousands of genetic differences (or mutations, usually Single Nucleotide Polyorphisms, or SNPs) identified in a patient's genome are likely to cause diseases. Although there are now many bioinformatics methods that try to solve this problem, they mostly work as follows. Information is collected about harmful mutations that are known to cause diseases and benign mutations that are not likely to cause diseases. Do they fall into protein coding regions? Do they affect splicing? Are they identified in Genome-wide association studies? Are they close to eQTLs? Do they fall in regions bound by proteins in Chromatin IP experiments? Do they fall into regions that are conserved over evolution? You get the idea. All of these types of information (called "features" in machine learning jargon) are then integrated in a model that predicts (or classifies) the known examples of harmful and benign mutations. Finally, newly discovered mutations can be fed into the computer (along with all the relevant features), and the model will predict whether the new mutation is harmful or benign.

We now turn to one of the major applications of machine learning in modern molecular biology and bioinformatics. Classification is one of the most important areas of machine learning in general, because it touches on what we really want when we say we want a machine to learn. For example, on the Internet, we want computers to tell the difference between photos of apples and oranges, cars, trucks and bikes, faces of friends and strangers.

In molecular biology, we seek to use genomic data to distinguish between driver and passenger mutations; between housekeeping genes and master regulators; between alpha helices and beta sheets; between transmembrane, cytosolic, nuclear or mitochondrial proteins; between T cells, B cells and stem cells; between Neanderthal and modern human DNA. All of these are classification tasks.

CHAPTER 10

Linear Classification

It's important to say at the outset that since classification is one of the most important and well-studied areas in machine learning, the number and diversity of classification methods is very large. These range from things you've probably heard of like Neural Nets and naïve Bayes to obscure sounding things like AdaBoost with decision stumps. We won't even begin to cover them all here. I will try to focus on a few methods that are easily connected to regression and clustering and that are widely used in molecular biology applications.

WIDELY-USED BIOINFORMATICS SOFTWARE
THAT IS DOING CLASSIFICATION

- PSIPRED (Jones 1999) classifies between secondary structures (alpha helix, beta sheet, etc.).
- PSORT (Horton et al. 2007) classifies between subcellular localization of proteins (nucleus, membrane, etc.).
- PolyPhen2 (Adzhubei et al. 2010) classifies between different types of SNPs (deleterious or benign).
- Pfam (Finn et al. 2014) classifies proteins into domain families (thousands of families representing conserved modular protein domains).

Most simply, we can think of a machine learning classification method as a sniffer dog at the airport. Its job is to tell us which suitcases contain illegal substances and leave the other suitcases alone. The sniffer dog has been trained beforehand with examples of both kinds of suitcases; we hope that during training, the sniffer dog smelled thousands of different smells, and learned to recognize which odorants are associated with illegal suitcases and which odorants are associated with suitcases that should be allowed to pass through. If the sniffer dog is doing its job well, we want it to bark when there really are illegal substances in new suitcases, and to stay silent when suitcases are fine to pass through.

The sniffer-dog example illustrates that real-world classification problems almost always involve high-dimensional data, and during training, good classification methods should be able to pick out which are the informative dimensions (features) for the task. For illustrative purposes, we will mostly stick with two-dimensional examples, but remember that we usually have high-dimensional problems in mind, and we do not know in advance which of these features are informative. We have already seen this type of feature-selection problem in linear regression, but it becomes much more important in classification. In classification, we are not interested in how well the model can explain the observations we have already seen—now we want to be able to make predictions about new examples. If we have included a bunch of useless or distracting features (or dimensions) in the model because they fit some of the noise in the training data, these will come back to bite us (much like a poorly trained sniffer dog) if distractions appear again in the new examples. In classification problems, overfitting is a serious practical concern. We'll spend Chapter 12 discussing the evaluation of classifiers.

COMBATING OVERFITTING IN CLASSIFICATION:
THE IDEAL CLASSIFICATION SETUP

1. Estimate the parameters of the classifier using known examples.
2. Determine the best parameters and choose the best classification method using a "validation set" of additional known examples.
3. See how your selected classifier does on an unseen "test" dataset of even more known examples.

Steps 1 and 2 are typically referred to as "training" and step 3 is referred to as "testing." In steps 2 and 3, a variety of approaches can be used to determine how well the classification method is working. We'll return to the issues of evaluating classifiers in Chapter 12.

10.1 CLASSIFICATION BOUNDARIES AND LINEAR CLASSIFICATION

All classification methods aim to find a boundary between groups of data-points that separates the different classes in the high-dimensional space of observations, often referred to as the "feature space." Linear classification methods aim to find a boundary between classes that can be expressed as a linear function of the observations in each dimension (often referred to as the "features"). For example, let's consider a typical molecular biology task of classifying cell types based on gene expression patterns; in this case, we'll try to classify T cells from other cells based on gene expression data from ImmGen (Figure 10.1).

Notice that one of the linear classification boundaries corresponds to a horizontal line. This means that only the data on the vertical axis are being used for classification—a one-dimensional classification. This is equivalent to simply ranking the cells by one of the expression levels and choosing a cutoff—above the cutoff we say everything is a T cell. I hope it's clear that by using both dimensions (a diagonal line) we'll be able to do a slightly better job of separating the T cells (+) from all others. You can see that even though the dimensions are highly correlated, there is still a lot of information gained by using both of them. As we go from a 1-D to a 2-D classifier, we'll have to train an additional parameter to write the equation for the linear classification boundary as opposed to simple

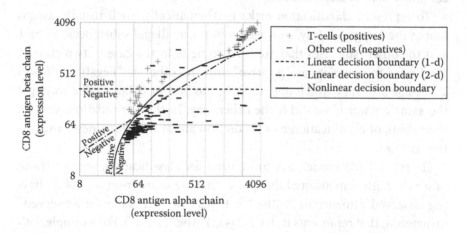

FIGURE 10.1 Decision boundaries in a two-dimensional feature space. Possible decision boundaries for classification of T cells based on gene expression levels of the CD8 alpha and beta chain.

cutoff. All linear classification methods will draw a line (or plane in higher dimensions) through the data, although they will disagree about what the line should be.

Of course, we might be able to do better by choosing a curve (or more generally a nonlinear boundary) to separate the classes. In this example, we can reduce the false positives (negatives on the positive side of the boundary; we will return to evaluation of classifiers in Chapter 12) by using an ellipse as the classification boundary instead of a line. However, as we go to more complicated high-dimensional nonlinear functions, we'll need more and more parameters to specify the classification boundary. We will return to nonlinear classification boundaries, so-called nonlinear classification, in Chapter 11. In general, in designing machine-learning methods for classification, we will face a trade-off between choosing the most accurate, possibly complex classification boundary versus a simpler (fewest parameters and dimensions) boundary whose parameters we can actually train.

10.2 PROBABILISTIC CLASSIFICATION MODELS

Luckily for us, several widely used classification methods follow directly from the probabilistic models I described for linear regression and clustering. For these classification methods, we'll have the likelihood as the objective function and train parameters by maximizing the likelihood. We'll therefore focus on these familiar models to illustrate the mathematical magic behind classification.

To represent classification tasks mathematically, we'll usually assign one of the categories (say, suitcases containing illegal substances) to be 1 and the other (suitcases that are fine) to be 0. In the case of two classes, we'll refer to one class as "positive" and the other as "negative." It will always be arbitrary which one we assign to be 1 or 0, but we'll always get the same answer if we did it the other way. In this formulation, we can then think of classification as a prediction about whether a new observation is 1 or 0.

If probabilistic models are to be used for classification, we can think of a new high-dimensional datapoint that we draw from a pool as having observed dimensions X (the "features"), as well as an "unobserved" dimension, that represents its true class, Y (the "target"). For example, let's say we have trained a sniffer dog (or classification model) and we are given a new suitcase (or observation). We observe X_{n+1} and we wish to fill in the unobserved corresponding Y_{n+1} based on the training data $X_1, ..., X_n$ and $Y_1, ..., Y_n$. If we are using a probabilistic model, we will use the training data to estimate some parameters θ, for example, by maximizing the likelihood $P(X_1, \ldots, X_n, Y_1, \ldots, Y_n|\theta)$. In addition to choosing the parameters of the model to maximize an objective function (typically done in the training stage), we will therefore also have to consider a rule to fill in the "unobserved" data Y_{n+1} (in the prediction stage). Since the objective functions for the training probabilistic models will be the usual suspects that we've seen for regression and clustering (likelihood, posterior probability, penalized likelihood, etc.), we won't discuss them again here. Instead, we'll spend a lot of time thinking about how to decide (mathematically) whether the new suitcase, number $n + 1$ (with associated features X_{n+1}), should pass through (whether the unobserved Y_{n+1} was really 1 or 0).

The maximum likelihood classification rule says (1) calculate the probability of observing the smells of that suitcase given that there *are* illegal substances in the suitcase, and the probability of observing the same smells given that there *are no* illegal substances, and (2) bark (assign the suitcase to the positive class) if the probability of the smell given that there *are* illegal substances is higher. In other words, the rule says assign X_{n+1} to class k such that $P(X_{n+1}|Y_{n+1} = k, \theta)$ is maximized.

On the other hand, the MAP classification rule says: (1) Calculate the posterior probability that there *are* illegal substances in the suitcase given the smells that were observed. (2) Bark (assign the suitcase to the positive class) if the posterior probability is greater than 0.5 (since there are only two choices). In other words, assign X_{n+1} to class k such that $P(Y_{n+1} = k|X_{n+1}, \theta)$ is maximized.

As you can imagine, other classification rules are possible for probabilistic classification models and absolutely necessary for classifiers that don't have a probabilistic interpretation. Furthermore, even if a classifier has a probabilistic interpretation, it might not be feasible to compute the likelihood or the posterior probabilities, so another rule will be needed. A simple example of a nonprobabilistic classification rule is "nearest neighbour" classification. It says—assign the new observation to the class of the observation that is closest in the training set.

SOME POPULAR CLASSIFICATION RULES

ML: Assign to class k, such that $P(X_{n+1}|Y_{n+1} = k)$ is maximized.
MAP: Assign to class k, such that $P(Y_{n+1} = k|X_{n+1})$ is maximized.
Nearest-neighbor: Assign to Y_i, such that $d(X_{n+1}, X_i)$ is minimized.

10.3 LOGISTIC REGRESSION

In Chapter 7, I gave the example of logistic regression: predicting whether something (Y) was 1 or 0 based on a series of observations (X). Thus, logistic regression can be thought of a classification method, and regularized multinomial regression is a respectable modern machine learning approach to multi-class classification. The estimation of the logistic regression parameters, b, is the "training" stage for this classification method. We assume that we are in a multivariate setting, so that we have a vector b with one component for each dimension, and we can proceed with estimation using one of the approaches we saw in the chapters on regression. To predict Y_{n+1} (or do the classification) based on the features, X_{n+1}, for a new observation we simply follow the assumption of logistic regression, and calculate

$$P(Y_{n+1} = 1 | X_{n+1}) = \frac{1}{1 + e^{-bX_{n+1}}}$$

Since there are only two possibilities (1 or 0), if this is greater than ½, we can assign observation $n + 1$ to the positive class by the MAP rule. The classification boundary is therefore

$$\frac{1}{2} = \frac{1}{1 + e^{-bX}}$$

which works out to $bX = 0$ (see Exercises). More generally, we can choose a threshold, t, and set the classification boundary to be $bX = t$, above which we will assign observation $n + 1$ to the positive class. This formula is just the equation for a line if there are two dimensions, i.e., $X_i = (X_{i1}, X_{i2})$, (see Exercises), and a plane (or "hyperplane"), if there are more than two dimensions.

GENE EXPRESSION SIGNATURES AND MOLECULAR DIAGNOSTICS

I showed in the earlier figure that it's possible to distinguish T cells from all other cells reasonably well based only on CD8 expression levels. Identification of combinations of genes whose expression levels can reliably distinguish cells of different types is of substantial medical relevance. For example, given a biopsy or patient's tumor sample, doctors would like to know (1) what types of cells are found in the sample, and more importantly, (2) do these cells represent a specific disease subtype? If so, (3) is this patient's disease likely to respond to a certain treatment or therapy? All of these are classification questions, and a large research effort has been directed toward finding specific classifiers for each of these challenges.

In fact, there are probably *many* combinations of genes whose expression levels distinguish between cell types. Once again, this represents a difficult feature selection problem: We have ~25,000 genes that can all be possible predictors, and we'd like to choose genes or combinations of genes that are the best. And the number of combinations is astronomical—even for 2 gene combinations there are more than 300 million possible predictors. Previously we saw that regularization can be used to choose sparse predictors in linear regression. What about for classification? Well, since logistic regression *is* linear regression, we can go ahead and use the same regularization strategy that we used before to obtain sparse classifiers.

Using logistic regression with $L1$ regularization and $\lambda = 0.25$, I obtained a two gene combination that does an almost perfect job at separating T cells from all other cells (Figure 10.2): only one false positive (other cell on the T-cell side of the boundary) and one false negative (one T cell on the other side of the boundary). Note that this problem would have been impossible with standard logistic regression (without the penalty) because the number of genes (dimensions of X) is much greater than the number of cells (the number of observations). Impressively, one of the genes chosen by the model is the T-cell receptor, whose expression level can identify T cells reasonably well on its own—but the computer didn't know this was the T-cell receptor when it chose it.

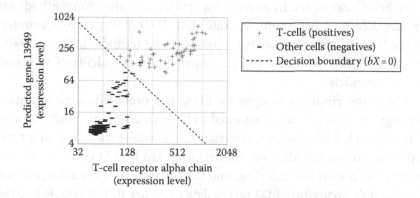

FIGURE 10.2   Identification of a two-gene signature using $L1$ regularized logistic regression. The computer automatically identified 2 genes (out of millions of possible two-gene combinations) that almost perfectly separates T-cells from other cells. One of these genes turns out to be the T-cell receptor.

Identifying gene signatures that predict clinical outcomes is one of the most promising avenues for genomics research. However, there are many complex statistical issues that need to be considered—methods to reliably identify these signatures is an area of current research interest (Majewski and Bernards 2011).

10.4 LINEAR DISCRIMINANT ANALYSIS (LDA) AND THE LOG LIKELIHOOD RATIO

In Chapter 6, we considered clustering using "hidden variables" that were 1 if the datapoint was in a particular cluster, and 0 otherwise. We showed that the computer could automatically learn a different model for each cluster or hidden state. The algorithm would then automatically separate datapoints into clusters corresponding to the values of these hidden variables.

If we think of the two "clusters" in the data as the two classes that we would like to separate (e.g., the suitcases with illegal substances and those that should be allowed to pass through), clustering using probabilistic models with hidden variables could be considered a classification task, except that we didn't do any training to learn what each type of suitcases smelled like. Clustering is therefore also called "unsupervised classification" to reflect that it can be thought of as classification without training (or supervision, in the machine learning jargon). Standard classification is referred to as "supervised" because we first teach the algorithm (sniffer dog) what the classes look like by showing it "training data" examples that are "labeled" by their class, and then apply it on new "unlabeled" examples. In general, supervised classification methods are expected to be much more accurate than their unsupervised counterparts. We will now consider two classification methods (LDA and Naïve Bayes) that can be considered the supervised equivalents of Gaussian mixture models.

Linear discriminant analysis (or LDA) is a probabilistic classification strategy where the data are assumed to have Gaussian distributions with different means but the same covariance, and where classification is typically done using the ML rule. The training step for LDA consists of estimating the means for each class, and the covariance of the entire dataset. For example, to perform LDA on the data in Figure 10.1, we would find the mean expression for the T cells and the mean expression for all other cells, say μ_1 (the mean when $Y = 1$) and μ_0 (the mean when $Y = 0$). We then estimate a single covariance matrix Σ for the whole data. This is done using the standard formulas for multivariate Gaussians that we saw in Chapter 4. For a new datapoint number $n + 1$ (dropping the $n + 1$ subscript notation for simplicity), we evaluate the ML rule

$$P(X|\mu_1,\Sigma) > P(X|\mu_0,\Sigma)$$

for the specific case of LDA, or more generally

$$P(X|Y = 1) > P(X|Y = 0)$$

for binary classification. Dividing both sides by the probability of the data given the negative class, we have

$$\frac{P(X|Y = 1)}{P(X|Y = 0)} > 1$$

The quantity on the left of this equation can be thought of as a likelihood ratio (LR), because we compute the probability of the data (a single observation, X) given the model for the positive class and compare it to the probability of the same data given the model for the negative class. Thus, the ML rule can be written as $LR > 1$, or more often as

$$\log LR = \log\frac{P(X|Y = 1)}{P(X|Y = 0)} > 0$$

The log-likelihood ratio (log LR) is a widely used statistic for classification. (It should not be confused with the likelihood ratio test statistic, and does not, in general, have a known distribution under a null hypothesis.) Going back to the specific case of LDA where we assume Gaussian models, this is

$$\log\frac{N(X|\mu_1,\Sigma)}{N(X|\mu_0,\Sigma)} > 0$$

which can actually be solved analytically to be

$$X^T\Sigma^{-1}(\mu_1 - \mu_0) > \frac{(\mu_1 + \mu_0)^T}{2}\Sigma^{-1}(\mu_1 - \mu_0)$$

Although this formula is a bit complicated, it actually has a very elegant geometric interpretation. It's easiest to see this by first considering the case where the covariance is the identity matrix. If the covariance is I,

$$X^T (\mu_1 - \mu_0) > \frac{(\mu_1 + \mu_0)^T}{2} (\mu_1 - \mu_0)$$

The left side of this equation says to take the new observation X and project it onto the vector that is the difference between the two means. The right is the projection of the average of the two means onto the difference of the two means. Geometrically, this is the midpoint along the vector that connects the two means. So the ML rule says to classify the observation as a positive if the projection of the observation onto the difference is more than halfway along the line between the two means (illustrated in Figure 10.3).

I hope this makes a lot of sense: In the LDA model, where the variance is the same in the two classes, the vector between the two means is the direction that contains the information about difference between the two classes. Therefore, the projection on this vector is the part of the observation that contains the information about which class the new observation is in. We can think about LDA as a projection of the observation vector onto a one-dimensional coordinate system that optimally separates the two classes. In that dimension, you simply assign the new observation to the mean that it's closer to (more than halfway in the two class scenario).

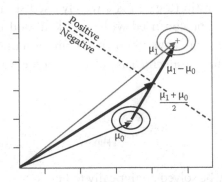

FIGURE 10.3 Geometric interpretation of the LDA classification rule. Vectors μ_1 and μ_0 illustrate the means for the positive and negative classes respectively. The decision boundary (dotted line) is orthogonal to the vector between the two means ($\mu_1 - \mu_0$) and intersects at the average of the two means ($\mu1 + \mu2$)/2.

All the other components of X can be thought of as distractions: They don't tell us anything about which class X belongs to.

This geometric interpretation also tells us how to think about the general case where the covariance of the data is not the identity: Before we project the observation X onto the difference between the means, we do another transformation of the space. This corresponds to a *rotation* or multiplication by the inverse of the covariance matrix. Indeed, common practice in linear classification is to estimate a single covariance matrix and multiply the data by the inverse of that matrix (which corresponds to a rotation). This transforms the data into a coordinate system where the covariance is I, and is called "whitening" because white noise is considered to be a Gaussian with covariance I.

DERIVING THE LDA DECISION BOUNDARY

We start with the ML rule, namely, assign the data, X, to the more likely class:

$$\log \frac{N(X \mid \mu_1, \Sigma)}{N(X \mid \mu_0, \Sigma)} > 0$$

To derive the decision boundary, let's first look at the log of the Gaussian distribution.

$$\log N(X \mid \mu, \Sigma) = \log \left[\frac{1}{\sqrt{|\Sigma|(2\pi)^d}} e^{-(1/2)(X-\mu)^T \Sigma^{-1}(X-\mu)} \right]$$

$$= -\frac{1}{2} \log \left[|\Sigma|(2\pi)^d \right] - \frac{1}{2}(X-\mu)^T \Sigma^{-1}(X-\mu)$$

Notice that the first term depends only on the covariance (and the dimensionality, d) which, by assumption, is the same for both classes in LDA. This means that this term will cancel out in the log ratio. Multiplying out the second term, we get

$$-\frac{1}{2}(X-\mu)^T \Sigma^{-1}(X-\mu) = -\frac{1}{2} \left[X^T \Sigma^{-1} X - 2X^T \Sigma^{-1} \mu + \mu^T \Sigma^{-1} \mu \right]$$

and, once again, we have one term that only depends on the covariance and the data. This term will therefore cancel out. We will be left with

$$\log \frac{N(X \mid \mu_1, \Sigma)}{N(X \mid \mu_0, \Sigma)} = -\frac{1}{2} \left[-2X^T \Sigma^{-1} \mu_1 + \mu_1^T \Sigma^{-1} \mu_1 + 2X^T \Sigma^{-1} \mu_0 - \mu_0^T \Sigma^{-1} \mu_0 \right] > 0$$

We now factor the terms and multiply by ½.

$$\log \frac{N(X \mid \mu_1, \Sigma)}{N(X \mid \mu_0, \Sigma)} = X^T \Sigma^{-1}(\mu_1 + \mu_0) + \frac{1}{2}(\mu_1 + \mu_0)^T \Sigma^{-1}(\mu_1 - \mu_0) > 0$$

which is the formula given above.

10.5 GENERATIVE AND DISCRIMINATIVE MODELS FOR CLASSIFICATION

One way to make sense of different types of classification strategies is to consider whether the classification method tries to make a probabilistic model for X (i.e., $P(X|\theta)$), or whether it models $P(Y|X)$ directly (Figure 10.4). In the machine learning jargon, these two types of methods are referred to as generative and discriminative, respectively. For the methods described so far, LDA is a generative method because it assumes a multivariate Gaussian model for each class. Generative models are appealing because it's easy to state the assumptions and to understand what the data would look like under the ideal case. Generative methods also tend to work better when the models they assume about the data are reasonably realistic, but they can also work pretty well even in certain situations when the data doesn't fit the model all that well.

For example, the idea of whitening and then classifying makes sense for LDA because the two classes are assumed to have the same covariance.

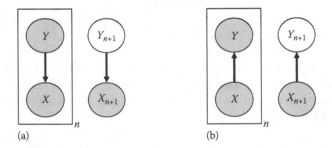

(a)   (b)

FIGURE 10.4 Structure of probabilistic classification models. Both generative and discriminative models for classification try to predict the unobserved class (Y) for a new datapoint ($n + 1$) based on a model trained on observed training examples. (a) The generative model defines the distribution of the features given the class, and then tries to infer the unobserved class using Bayes theorem. (b) The discriminative model assumes that the probability distribution of Y directly depends on the features, X.

This assumption also makes the classification much less sensitive to errors in estimation of the covariance. As I've mentioned a few times already, estimating the covariance for multivariate Gaussians is a major practical hurdle for high-dimensional data. However, because in LDA errors in the estimation of the covariance will be the same for both classes, the impact of bad estimation on the classification decision will tend to cancel out. In general, the models assumed by generative methods will not actually fit the data, so it's wise to use a classification strategy that will not assign new observations to one class because they simply fit the assumptions of the classifier better. However, if the data for both classes deviates from the model in the same (or similar) way, the generative classifier will often do just fine. For example, imagine that the new datapoint you observed was *very* far away from both mean vectors in LDA, and was extremely unlikely under the Gaussian models for both classes. You can still go ahead and classify this new observation because only the part of the observation that's in the direction of the difference between means is important. This is one nice feature of classification using a likelihood ratio: You don't need to claim that the model for either class fits the data very well—you can still assign the new data to the class that fits best.

Discriminative methods are classification strategies that don't try to make a model of the data for each class. They simply try to get a good classification boundary. Logistic regression is a great example of a discriminative strategy. In general, these methods will work better when the data fit very poorly to a generative model (or a generative model requires large numbers of parameters that are hard to estimate). The downside of discriminative models is that they are harder to interpret and understand— they are more of a "black box"—and this can make it more difficult to train parameters while avoiding overfitting. Since they don't actually make a model of the data, discriminative models don't use the ML rule.

10.6 NAÏVE BAYES: GENERATIVE MAP CLASSIFICATION

Naïve Bayes is one of the most widely used classification strategies and does surprisingly well in many practical situations. Naïve Bayes is a generative method and allows each dimension to have its own distribution. However, the key difference between naïve Bayes and LDA is that naïve Bayes assumes that the dimensions are *independent* (Figure 10.5). This way, even if the data are modeled as Gaussian, the number of parameters is linear in the number of dimensions. Finally, naïve Bayes uses the MAP classification rule so that it also includes a prior for each class, which

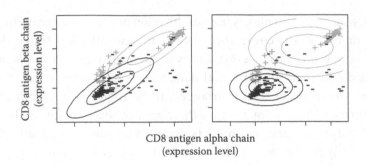

FIGURE 10.5 Different generative models underlie LDA and Naïve Bayes. This figure shows a comparison of the Gaussian models estimated for the two classification models. On the left, LDA assumes a single covariance matrix that captures the strong positive correlation in this data. On the right, naïve Bayes captures the greater variance in the positive class by using different variances for each class, but assumes the dimensions are independent, and therefore misses the covariance. In this example, neither class fits Gaussians very well, so neither generative classification model is expected to find the optimal (linear) classification boundary.

turns out to be very important for datasets where the number of positives and negatives are not similar. To use the MAP rule when there are two classes, we'd like to compare the posterior to 0.5, so the decision boundary is given by

$$P(Y = 1|X) = \frac{1}{2}$$

As before, X represents a single new observation, and Y represents the unobserved (predicted) class of that observation. Since we have a generative model, we use Bayes' theorem and write

$$P(Y = 1|X) = \frac{P(X|Y = 1)P(Y = 1)}{P(X)}$$

$$= \frac{P(X|Y = 1)P(Y = 1)}{P(X|Y = 1)P(Y = 1) + P(X|Y = 0)P(Y = 0)} = \frac{1}{2}$$

As I mentioned, the key assumption of the naïve Bayes classifier is that it assumes that all the dimensions of X are independent conditioned on the class. This means we can write

$$P(X|Y=1) = \prod_{j=1}^{m} P(X_j|\theta_{1j}, Y=1)$$

where I have written θ_1 for the parameters of the probability distribution (or model) for the data of each dimension for the positive class. This formula says: To get the probability of the multidimensional observation, X, we just multiply together the probabilities of each component of X. Of course, we still have to define the probability distribution for each dimension for each class, but I hope it's clear that the assumption of independence between dimensions leads to a great simplification. For example, if we choose a Gaussian distribution, we don't have to worry about covariance matrices: We simply need a single mean and variance for each class for each dimension.

Plugging the independence assumption back into the formula for the posterior probability gives

$$P(Y=1|X) = \frac{1}{1+((1-\pi)/\pi)e^{-\sum_{j=1}^{m}\log(p(X_j|Y=1)/p(X_j|Y=0))}} = \frac{1}{2}$$

where I have used π to represent the prior probability of observing the positive class. Under Naïve Bayes, the decision boundary turns out to be a logistic function based on the sum of the log likelihood ratios for each class, with an extra parameter related to the priors. This equation can be solved easily just by thinking about when the logistic function equals exactly ½.

$$\frac{(1-\pi)}{\pi}e^{-\sum_{j=1}^{m}\log(p(X_j|Y=1)/p(X_j|Y=0))} = 1$$

or

$$\sum_{j=1}^{m}\log\frac{p(X_j|Y=1)}{p(X_j|Y=0)} = \log\frac{(1-\pi)}{\pi}$$

So the MAP rule says to compare the log-likelihood ratio to a cutoff related to the priors: The smaller the prior on the positive class, the larger the likelihood ratio (which is based on the data) needs to be before you classify the new datapoint as positive. In this way, the priors allow us to represent the maxim "extraordinary claims require extraordinary evidence."

We can go ahead and make this more specific by assuming distributions for each dimension. For example, in the case of categorical distributions for each dimension (as used for sequence data in Chapters 4 and 6), we can use clever indicator variable notation from Chapter 6 to write

$$P(X|f, Y = k) = \prod_{j=1}^{m} \prod_{b} f_{kjb}^{X_{jb}}$$

where f are the parameters of the discrete distribution for class k, corresponding to the categories b that could be observed in each dimension j. The naïve Bayes classification boundary works out to be

$$\sum_{j=1}^{m} \log \frac{p(X_j | Y = 1)}{p(X_j | Y = 0)} = \sum_{j=1}^{m} \sum_{b} X_j \log \frac{f_{1jb}}{f_{0jb}} = \log \frac{(1 - \pi)}{\pi}$$

In fact, this is exactly the statistic used to identify new examples of DNA motifs in the genome.

IDENTIFYING MOTIF MATCHES IN DNA SEQUENCES

A classic bioinformatics problem is to find examples (sometimes called instances) of weak patterns (called motifs) in DNA (or protein) sequences. For example, based on known positive examples (training set), we estimate a model for the specificity of a transcription factor. This model gives the probability of observing DNA bases at each position in the known positives. So if we have a pattern of width w, we will have a matrix whose entries give the probability of observing each of the four bases (A, C, G, and T) at each of the w positions in the motif. Training examples of known DNA binding sites are shown in Table 10.1.

We then go ahead and estimate the parameters for the motif probability matrix:

$$f = \begin{matrix} 0.083 & 0.667 & 0.083 & 0.75 & 0.667 & 0.167 \\ 0.083 & 0.083 & 0.083 & 0.083 & 0.083 & 0.25 \\ 0.75 & 0.167 & 0.083 & 0.083 & 0.167 & 0.417 \\ 0.083 & 0.083 & 0.75 & 0.083 & 0.083 & 0.167 \end{matrix}$$

Given the motif model, we want to find new examples of the binding site in a long DNA sequence. To think of this as a classification problem (illustrated

TABLE 10.1   Binding Sites from Promoters of GATA
Regulated Genes in the SCPD Database

Gene	Binding Site
YIR032C	GATAAG
YIR032C	GGTAAG
YIR032C	GATAAG
YJL110C	GATAAT
YKR034W	GATAGA
YKR034W	GATAAC
YKR039W	GATAAG
YKR039W	GATAAC

Source: Zhu, J. and Zhang, M.Q., *Bioinformatics*, 15(7–8),
607, 1999.

X = GTTGATAACGAGTTTCCACCTTATCACTTATCACTAGTGCTAATCAAACAGCAAAGAATGCTTGATAGA A
Y = ??

X_1 GTTGAT

X_2 TTGATA

X_3 TGATAA

    GATAAC

    ATAACG

       ...

X_{n-w-1} TGATAG

X_{n-w} GATAGA

X_{n-w+1} ATAGAA

FIGURE 10.6   Motif matching as a classification problem. At each position in
the DNA sequence, we have a w-dimensional datapoint ($w = 6$ in this example).
Our goal is to classify whether each position is an example of the motif or not. In
the naïve Bayes framework, this is done by computing $P(Y|X)$, assuming that the
dimensions (the six positions in the motif) are independent.

in Figure 10.6), our goal is to classify each subsequence of length w in a long
sequence as either an example or instance of this motif (positive) or a background sequence (negative).

Like many classification problems in bioinformatics, this training set consists only of positive examples of known binding sites. We are not given
any sequences where we know there is no binding by the transcription factor. However, we can use the biological intuition that the vast majority of
genomic sequences will not be bound by the transcription factor and assign
the negative class a "background model," say, g, which we assume has the
genome average probability of A, C, G, and T at each position.

$$g = \begin{matrix} 0.3 & 0.3 & 0.3 & 0.3 & 0.3 & 0.3 \\ 0.2 & 0.2 & 0.2 & 0.2 & 0.2 & 0.2 \\ 0.2 & 0.2 & 0.2 & 0.2 & 0.2 & 0.2 \\ 0.3 & 03 & 0.3 & 0.3 & 0.3 & 0.3 \end{matrix}$$

To classify a new position i as a motif instance or not, we calculate the likelihood of the sequence X_i under the motif model (as in Chapter 6) for the positive class and under the background model for the negative class

$$P(X_i|f) = \prod_{j=1}^{w} \prod_{b \in \{A,C,G,T\}} f_{jb}^{X_{ijb}}$$

$$P(X_i|g) = \prod_{j=1}^{w} \prod_{b \in \{A,C,G,T\}} g_{jb}^{X_{ijb}}$$

To derive the naïve Bayes classification boundary, we need to take the log ratio of these and compare to the ratio of the priors:

$$\sum_{j=1}^{m} \sum_{b \in \{A,C,G,T\}} X_j \log \frac{f_{jb}}{g_{jb}} = \log \frac{(1-\pi)}{\pi}$$

In practice, the log ratio of the priors is often referred to as a "cutoff" or "threshold," say, t, and typically a different cutoff is chosen for each motif model, f. We then identify motif matches at each sequence in the genome where

$$\sum_{j=1}^{m} \sum_{b \in \{A,C,G,T\}} X_j \log \frac{f_{jb}}{g_{jb}} > t$$

For ease of interpretation, the cutoff, t, can always be converted back into the prior using this formula.

$$\pi = \frac{1}{e^t + 1}$$

For example, a cutoff of 4 corresponds to a prior of $(1/e^4 + 1) = 0.018$. If we search DNA with a cutoff of 4, this means that (before seeing the data) we believe that less than 2% of the positions that we test correspond to examples of the motif.

Because this model is typically used to scan very long DNA sequences, the log ratio of parameters for each base b at each position j is often precomputed and stored as a "position specific scoring matrix" or PSSM. This matrix has the form

$$M_{jb} = \log \frac{f_{jb}}{g_{jb}}$$

When written this way, the naïve Bayes motif search procedure can be written as the (Frobenius) inner product of the PSSM with the datapoint. We should classify a position as a motif when

$$M \cdot X_i > t$$

10.7 TRAINING NAÏVE BAYES CLASSIFIERS

As with any classifier, during the training stage, we need to do estimation of the parameters for Naïve Bayes. For models for each class (i.e., Gaussians, Discrete, etc.) this is straightforward: simply take all of the known examples for each class and do ML estimation. Since naïve Bayes assumes all the dimensions are independent, this is always a simple univariate estimation problem for the known positives and negatives in each dimension. In practice, for categorical distributions, there are some subtleties here related to avoiding ML estimates of zero, but these are discussed elsewhere (e.g., Henikoff and Henikoff 1996).

We also need to estimate the prior parameters, which can turn out to be important. The ML estimate for the prior parameter of the positive class turns out to be simply the fraction of the training set in the positive class. However, using this estimate for classification of new observations means we are assuming that the fraction of positives in the new examples is the same as what it was in the training set. In practice, in molecular biology applications, this is not true. For example, let's say we were using a naïve Bayes classifier to recognize a certain feature in the genome, such as a DNA sequence motif. Ideally, we would train the classifier based on a set of positive examples and a set of examples that we know are not the DNA motif. In practice, we typically have a small number of positive examples, and *no* examples of sequences we know are not the motif. However, when we search the genome, we expect very few of the positions we test to be new examples of motifs. Even in the ideal case, the prior estimated based on the training data won't reflect our expectations for the situation when we make predictions. This strongly suggests that we should set the prior for classification of new examples to be much smaller that the fraction of positives in our training set.

10.8 NAÏVE BAYES AND DATA INTEGRATION

One of the most powerful consequences of assuming that dimensions are independent is that we can develop classification models that don't assume that all the dimensions follow the same probability model: We can use Poisson distributions for some dimensions, discrete distributions for others, Gaussians, Bernoulli, ... you get the idea. If we tried to model the correlation in two-dimensional data where one of the dimensions is 0, 1, 2, ..., and the other is any number between $-\infty$ and $+\infty$, we would have a very difficult time finding an appropriate distribution. Since naïve Bayes just ignores these correlations, it gives us a very simple and powerful way to combine data of different types into a single prediction.

A good example where naïve Bayes classification has been used to integrate multiple types of data to make predictions is protein–protein interactions (Jansen et al. 2003). Here, the task is to figure out which pairs of proteins physically interact in the cell based on a number of types of data, such as gene expression patterns, presence of sequence motifs, subcellular localization, or functional annotation. Since gene expression patterns can

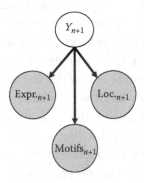

FIGURE 10.7 Graphical models representation of data integration. This simple generative model can be used to integrate different types of data. A naïve Bayes classifier can then be used to predict protein interactions based on multiple genomic measurements such as correlation of gene expression patterns (Expr.), shared subcellular localization (Loc.), and the presence of compatible motifs in the primary amino acid sequence (Motifs). The interaction for the new pair ($n + 1$) is assumed to be unobserved, but the data of each type is observed.

be systematically measured, this data is typically in the form of a real number. On the other hand, protein localizations and functional annotations are usually discrete categories. Sequence motifs might occur 0, 1, 2, ..., times, so they are natural numbers. If all these data are assumed to be independent, the parameters for each data type can be estimated independently, and all the features can be combined using a simple generative model (illustrated in Figure 10.7).

EXERCISES

1. Show that the decision boundary for logistic regression is actually $bX = 0$ as I claimed. Show that this is actually an equation for a line in two dimensions as I claimed.

2. The formula I gave for $P(X)$ in the MAP classification is only true for the two-class case. Write the more general formula for k-class classification.

3. Derive the decision boundary for naïve Bayes in the two-class Gaussian case. Show that in the case of uniform priors and equal covariance between the positive and negative classes, naïve Bayes is the same as LDA with diagonal covariance.

4. I said that discriminative classification methods don't use the ML rule. Make assumptions about priors to apply the ML rule to logistic regression. (*Hint*: Use Bayes theorem.)

5. Notice that the parameters I gave for the model, f, (in the DNA binding site example) were *not* the maximum likelihood estimates based on the positive training examples that I showed in the table. What are the MLEs for the positive class? Why didn't I use them for the classifier?

REFERNCES AND FURTHER READING

Adzhubei IA, Schmidt S, Peshkin L, Ramensky VE, Gerasimova A, Bork P, Kondrashov AS, Sunyaev SR. (2010). A method and server for predicting damaging missense mutations. *Nat. Methods* 7(4):248–249.

Finn RD, Bateman A, Clements J, Coggill P, Eberhardt RY, Eddy SR, Heger A et al. (2014). Pfam: The protein families database. *Nucleic Acids Res.* 42(Database issue):D222–D230.

Henikoff JG, Henikoff S. (1996). Using substitution probabilities to improve position-specific scoring matrices. *Comput. Appl. Biosci.* 12(2):135–143.

Horton P, Park K-J, Obayashi T, Fujita N, Harada H, Adams-Collier CJ, Nakai K. (2007). WoLF PSORT: Protein localization predictor. *Nucleic Acids Res.* 35(Web Server issue):W585–W587.

Jansen R, Yu H, Greenbaum D, Kluger Y, Krogan NJ, Chung S, Emili A, Snyder M, Greenblatt JF, Gerstein M. (2003). A Bayesian networks approach for predicting protein–protein interactions from genomic data. *Science* 302(5644):449–453.

Jones DT. (1999). Protein secondary structure prediction based on position-specific scoring matrices. *J. Mol. Biol.* 292(2):195–202.

Majewski IJ, Bernards R. (2011). Taming the dragon: Genomic biomarkers to individualize the treatment of cancer. *Nat. Med.* 17(3):304–312.

Zhu J, Zhang MQ. (1999). SCPD: A promoter database of the yeast *Saccharomyces cerevisiae. Bioinformatics* 15(7–8):607–611.

CHAPTER 11

Nonlinear Classification

IN THE PREVIOUS CHAPTER, we considered methods that draw a (high-dimensional) line through the feature space to separate classes. However, it's clear that for some datasets, there simply will be no line that does a very good job. In practice, it doesn't take much to get datasets where linear classification methods will not perform very well. However, once we accept that we aren't going to be able to use a line (or hyperplane) in the feature space to divide the classes, there are a lot of possible curves we could choose. Furthermore, there are only a few probabilistic models that will yield useful nonlinear classification boundaries, and these tend to have a lot of parameters, so they are not easy to apply in practice.

QDA: NONLINEAR GENERATIVE CLASSIFICATION

Quadratic discriminant analysis (or QDA) is the generalization of linear discriminant analysis (or LDA) to the case where the classes are described by multivariate Gaussian distributions, but they are not assumed to have equal covariances. QDA does not yield a linear classification boundary, and in principle would perform optimally in the case where the data truly were drawn from multivariate Gaussian distributions. In practice, this is rarely (never?) the case in molecular biology. Perhaps more problematic is that now we need to estimate covariance matrices for each class. As long as the number of true positive training examples is small (which is typically true for problems with rare positives), estimation of the covariance will be unreliable. Nevertheless, QDA is the lone well-studied example of a nonlinear generative classification model that's based on a standard

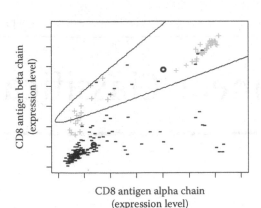

FIGURE 11.1 The decision boundary for QDA on the T-cell classification using CD8 gene expression is indicated by a solid line. The means for the two classes are indicated by black unfilled circles, and the datapoints are indicated as "+" for T-cells (positives) and "−" for other cells (negatives). In this case, the data doesn't fit the Gaussian model very well, so even though the classification boundary is nonlinear, it doesn't actually do any better than a linear boundary (it actually does a bit worse).

probability model (at least it's the only one I know about). The decision boundary turns out to be an ellipse (or quadratic form) in the feature space (Figure 11.1).

11.1 TWO APPROACHES TO CHOOSE NONLINEAR BOUNDARIES: DATA-GUIDED AND MULTIPLE SIMPLE UNITS

We will see two major strategies to nonlinear classification in the methods that follow. The first is to let the training data be a guide: we'll choose nonlinear classification boundaries based on the data that we have in the training set. This will allow us to come up with nontrivial nonlinear boundaries without having to estimate large numbers of parameters or consider obscure mathematical models. The main disadvantage is that the classifiers we obtain will always be dependent on the specific sample of data we observed for training. There will be no estimation of model parameters that give us a biological intuition or statistical model for our problem. The second major strategy is to build the nonlinear classifier by cumulating the effects of several simple linear classifiers. This way we can control the complexity of the classifier by only adding on additional simple linear classifiers when they improve the classification. The major drawback of these methods is that there is typically no obvious order or starting point for this process of building up the classifier. This randomness or contingency in the training procedure must be dealt with, adding extra complication to the procedure.

NEURAL NETWORKS: HOW DO WE (AND OTHER ANIMALS) DO CLASSIFICATION?

I've been discussing sniffer dogs as an analogy for a machine classification system, but we haven't thought about how animals actually manage to solve very complicated classification problems. Certainly, sniffer dogs in the airport are not doing LDA or any similar model-based calculation. As I write this, human nervous systems are still the most accurate and flexible classification methods known (expect this to change in the next few years). In the training or learning phase, it is believed that connections between neurons are modulated based on simple rules (such as Hebb's rule). Once the connections have been formed, sensory inputs are passed along through these connections ultimately leading to activation of a neuron (or group of neurons) whose specific activity represents the class of the observation.

Although understanding learned animal behavior is a complicated neurobiology problem, and outside of the scope of this book, animal nervous systems were the inspiration for a class of machine learning methods known as neural networks (or more accurately, artificial neural networks). Although we won't consider these further here, these methods fall into the category of nonlinear classification methods that build complex classifiers out of many simpler units. In the case of neural networks, the simple units are referred to as "neurons," and they are represented by simple mathematical functions. Because no probabilistic models underlie such networks, computer science researchers have developed objective functions for neural networks inspired by physical intuition (e.g., energy minimization, Poultney et al. 2006) and training methods based on biological analogy (e.g., wake–sleep algorithm, Hinton et al. 1995). An added level of complexity in these models is the choice of the structure of the neural network. Among the most practically successful artificial neural networks are those designed to recognize patterns in images, with structures inspired by the animal visual cortex (Le Cun et al. 1990). Finally, because large neural networks may have millions of parameters, learning the parameters requires specialized optimization techniques, and regularizing to prevent overfitting is a complex issue.

Where good solutions to these challenges have been identified, there has been considerable success using so-called "deep" neural networks to perform practical classification tasks (e.g., Krizhevsky et al. 2012). These models are referred to as "deep" because they have many layers (say more than three) of simple units ("neurons"). In deep neural networks (and deep learning), computations are done directly on the data by a first layer, and the results of those computations is sent to a second layer that does a further computation, the results of which are sent to a third layer, etc., such that information is propagated "deep" into the model. These models are designed to mimic the way sensory information is processed by layers of biological neurons in the retina and visual cortex. Because of their recent successes in a wide range of classification tasks, we should expect to see these methods increasingly applied to genome-scale biological data in the coming years.

11.2 DISTANCE-BASED CLASSIFICATION WITH k-NEAREST NEIGHBORS

A very simple (and therefore popular) nonlinear classification strategy is the "nearest neighbor" approach. The idea here is to define a distance in the high-dimensional feature space, and classify new observations based on the observations that we've already seen close to the location of the new observation in the feature space. In the simplest form, a somewhat formal description of the k-NN algorithm is as follows.

Earlier we defined the classification problem as follows. Given a new datapoint number $n + 1$, associated with features X_{n+1}, we wish to classify it in to one of j classes, represented by the unobserved vector Y_{n+1}, using a training set of $X_1, ..., X_n$ and observed classes $Y_1, ..., Y_n$.

With these definitions, the k-NN algorithm has two steps:

Step 1: For each datapoint in the training set, $X_1, ..., X_n$, compute the distance, $D_i = d(X_{n+1}, X_i)$. Define the k datapoints with the smallest distances (sort D_i and choose the first k) as the neighbors of datapoint $n + 1$.

Step 2: Count up the number of times each class appears among the neighbors of datapoint $n + 1$ and assign it to the most frequent class among the neighbors.

K-nearest neighbors can be thought of as the supervised analogue of hierarchical clustering, just as LDA can be thought of as the supervised version of k-means clustering. Starting from the new datapoint, we build a tree of the k most similar previously observed datapoints. Thus, we are assigning the new observation to a cluster of size k (the k most similar datapoints). Once we have identified the cluster or neighborhood of the unknown observation, we classify it based on what classes those datapoints fall into. I hope it's clear that this classification strategy can yield highly nonlinear and

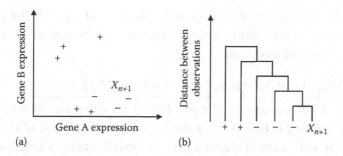

FIGURE 11.2 k-NN classification. (a) In k-nearest neighbor (k-NN) classification, the datapoint number $n + 1$, with features X_{n+1}, is assigned the class based on the classes of a fixed number of (k) neighbors. In this example, the majority of the nearest neighbors of the new datapoint are negatives, so the new data can be classified as a negative, even though the data are not linearly separable. (b) Nearest neighbor classification can be thought of as building a hierarchical tree of k neighbors based on the distances. Then, based on the frequency of each class in the "cluster" or neighborhood, the new datapoint is assigned a class.

discontinuous classification boundaries, and there is no assumption about the distribution of positive or negative datapoints. Perhaps most attractive about k-NN is that there is no dependence on the number of parameters as the dimensionality (or data size) grows, assuming that we can still compute the pairwise distances. In the example (Figure 11.2), if we use the five nearest neighbors ($k = 5$), we will classify the new datapoint as "negative."

An obvious problem with the k-NN approach is that (like K-means) we don't have a way of choosing k. However, since this is the only free parameter in the method, we can leave it for now and choose k that maximizes our performance evaluation measures (we will return to these in the Chapter 12).

Perhaps the most unsatisfying thing about nearest neighbor approaches to classification is that we need to keep the entire dataset of known examples as part of the classifier and compare the new observation to all the data we have seen before. From the perspective of a "predictive model," nearest neighbor classification is about the worst we can do: we haven't learned a compact representation or gained any scientific insight into the problem. It's the equivalent of predicting planetary orbits by identifying the most similar planets in the sky and using observations from those "neighboring" planets to predict the orbit of a new planet. No gravity needed. This reliance on searching the training set to make predictions also means that we need to keep the entire database of training data on hand and compute the distances of the new point with all the previous observations. If the training data is large, this can be slow and imply very large memory requirements.

11.3 SVMs FOR NONLINEAR CLASSIFICATION

Another clever way around the problem of choosing a nonlinear classification boundary is to look for a nonlinear transformation of the feature space that will make the data easier to classify using a linear decision boundary. This is illustrated in Figure 11.3 where the data in (a) have been transformed to the data in (b) using

$$X_3 = \sqrt{(X_1 - 6000)^2 + (X_2 - 3000)^2}$$

$$X_4 = \sqrt{(X_1 - 64)^2 + (X_2 - 256)^2}$$

And there's no reason to restrict the transformation to a space of the same dimensionality: it's possible that data that are not linearly separable in a two-dimensional feature space will be linearly separable in a higher dimensional space where the new dimensions are nonlinear functions of

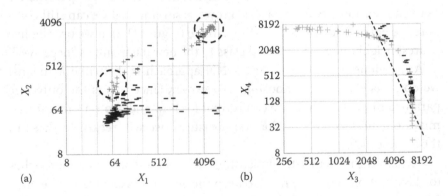

(a)

(b)

FIGURE 11.3 Nonlinear transformation followed by linear classification. (a) In a space where the coordinates are the expression levels of these two genes (X_1 and X_2), the T cells (positives, gray +) cannot be separated from the other cells (negatives, black −) using a line in the feature space. Two dotted circles represent a nonlinear transformation of the data that might improve separation. (b) Transformation of the data where the two new dimensions (X_3 and X_4) are the distances of each point from the center of the two dotted circles in (a). Now, a small value in either dimension is likely to be a positive example, and many of the T cells can be separated from the other cells using a linear classification boundary.

the features. Of course, the number of possible transformations is infinite, so it's hard to know if you've chosen the right (or best) one. Furthermore, a high-dimensional nonlinear transformation might have a lot of parameters to train (in this 2-D example, the centers of the circles are specified by four parameters). Fortunately, there is a trick that will allow us to solve both of these problems in practice: the kernel trick (yes, literally a trick). The idea of the kernel trick is to consider transformations that only depend on pairs of datapoints, just like the distances in k-NN. This will allow us to define complicated nonlinear transformations with very few parameters. Even more amazing, using kernels (think pairwise distances) will allow us to do classification in the transformed space without actually doing the calculations to transform the data. This is the kernel trick.

One of the most powerful and popular classification strategies that takes advantage of the kernel trick to do nonlinear classification is Support Vector Machines (or SVMs). SVMs are in fact a linear classification strategy that defines the classification boundary based on datapoints selected from the training set. It's easiest to explain this for the two-dimensional case, where the classification boundary is a line. In this case, the SVM chooses the datapoints (the "support vectors") that define the line that maximizes the region (in the feature space) between the positive and negative classes (the "margin"). To maximize the margin, the SVM finds the positive datapoints (support vectors) that define a line as far as possible from all the negative datapoints, and negative datapoints (more support vectors) that define a line as far as possible from all the positive data, with the constraint that these two lines are parallel (Figure 11.4a). Note that there is no likelihood or other probabilistic model being optimized here. The objective function of the SVM is simply the size of the separation between the classes, and all possible datapoints can be explored as possible support vectors. Unfortunately, there is no simple closed-form solution to this problem, but much like logistic regression, it can be solved reasonably reliably using numerical methods.

When the classes cannot be perfectly separated by a line (the data are not "linearly separable"), the points on the other side of the line (or within the margin) are included in the optimization problem, so that the objective function is now based on both the size of the margin *and* the number of points that are on the wrong side of the classification boundary (Figure 11.4b). In the simplest case, there is a constant penalty given to each point on the wrong side of the classification boundary, and this must be chosen by the user.

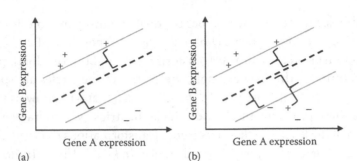

FIGURE 11.4 Support vector machines optimize the space between the positive and negative classes. Positive examples are indicated as "+" and negative examples are indicated as "–." (a) In this illustration, the support vectors are the points touching the gray lines. If the data are linearly separable, the SVM maximizes the distance between the positive and negative support vectors (indicated by braces) by choosing the decision boundary parallel to the hyperplanes defined by the support vectors. (b) If the data are not linearly separable, a penalty is added to the objective function for each point on the wrong side of the decision boundary (indicated by an additional brace).

As I'm sure you have already guessed, it's straightforward to generalize all of this to an arbitrary number of dimensions where lines become "hyperplanes." Amazingly (and probably the reason for the success of SVMs in practical high-dimensional problems), everything else simply follows. Because SVMs use the data to choose the decision boundary (and not fancy probabilistic models), the number of parameters will not necessarily increase with the dimensionality of the data. This makes them less prone to overfitting high-dimensional data. Unlike k-NN that needs to store the entire dataset to make the next prediction, SVMs only need to store the "support vectors"—the few datapoints at the margins. Since the number of support vectors used is typically a user-controlled parameter, the memory requirements of SVMs can be kept reasonable.

Although SVMs are a good linear classification strategy, this chapter is about nonlinear classification. And the reason I have spent so much time explaining SVMs is the following: it turns out that the decision boundary for SVMs can be expressed (mathematically) as a function of the dot products between the datapoints and the "support vectors" (which are nothing more than the datapoints chosen by the algorithm to use for the classification boundary). This means that SVMs can be generalized to nonlinear classification by thinking about the dot product as a type of distance (technically it's a similarity, not a distance, but see the discussion in Chapter 5). Thus, we can think of SVMs as a kind of "similarity-based" classification method (more technically, a "kernel"-based classification method).

Amazingly, this means that it's possible to run the exact same algorithms to train SVMs using different kernels (or pairwise similarity metrics). Thus, it's possible to do a nonlinear transformation of the data by selecting a nonlinear kernel, and then apply SVMs to do linear classification on the transformed data (in the transformed, possibly higher dimensional space). Because the kernels depend only on the pairs of datapoints, any nonlinear transformation that can be expressed as a pairwise similarity measure (as a kernel) can be used for classification. Furthermore, because the nonlinearity only appears in the kernel, it's not actually necessary to compute the nonlinear transformation. It's difficult to overstate the importance of this result: in practice, this means that SVMs can perform very well on a wide variety of classification problems where the data are not linearly separable in the feature space. Figure 11.5 shows a comparison of a linear and nonlinear SVM on the T-cell classification example.

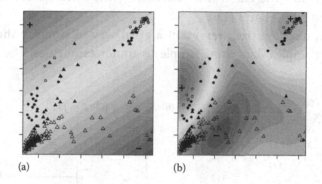

(a)   (b)

FIGURE 11.5   Classification of T cells (as in Figure 11.1) using an SVM. In this representation (made using the R package kernlab, Karatzoglou et al. 2016), darker gray areas are more confidently assigned to one class (indicated with a + for positive class and – for the negative class), whereas the lightest areas are the classification boundaries. The positive class is indicated by circles, while the negative class is indicated by triangles. Filled symbols indicate the datapoints chosen as the "support vectors" that define the classification boundaries. (a) A linear kernel leads to a linear classification boundary and cannot separate the classes very well. (b) A Gaussian kernel (or RBF, radial basis function) leads to a highly nonlinear (and discontinuous) classification boundary that correctly identifies the two groups of positives and allows negatives to appear between them.

In general, however, we don't know which kernel will perform well on our data, and this needs to be determined by experimentation—as far as I know, there's no good way to visualize the high-dimensional data to get a sense of what kind of transformation will work. And kernels have parameters. In practice, however, kernels with relatively few parameters that correspond to projections to high-dimensional spaces work very well. Furthermore, because they're relatively straightforward to compute, if you can try one, you can usually try a few.

11.4 DECISION TREES

Another way to develop a nonlinear classification boundary is to use several linear classifiers to approximate the nonlinear one. By applying simple classifiers sequentially, it's possible to build up a very accurate nonlinear one. "Decision trees" are one of the simplest ways to do this.

Decision trees not only use linear classifiers at each step, but they restrict these to be univariate (one-dimensional). Thus, a decision tree corresponds to a series of simple "cutoffs" at each step. These cutoffs are arranged hierarchically, and because at each stage a datapoint be either above or below the cutoff (the "decision"), the series of cutoffs form a bifurcating tree (Figure 11.6).

Clearly, decision trees represent a great way to simplify the classification of complex data using simple-to-understand rules. The problem,

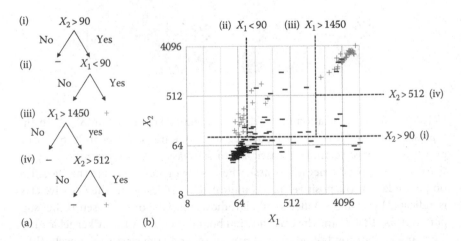

FIGURE 11.6 Decision trees. (a) An example of a four-layer decision tree to classify the data in (b). In (a), the numbers (i)–(iv) correspond to the decision boundaries (dotted lines) in (b).

of course, is how we chose the rules—in other words, how do we train the classifier? The classical approach to do this is an algorithm known as CART (Breiman et al. 1984). The idea is to choose the cutoff at each stage that maximizes the "purity" of the training data on either side of the cutoff. There are different ways to measure the "purity," but in the popular CART algorithm, it was done by minimizing the "Gini impurity," which is

$$I = \sum_k f_k(1 - f_k)$$

where
   k indexes the classes (typically positive or negative)
   f is the fraction of the datapoints in the class

So at each step the CART algorithm chooses a cutoff in one dimension by minimizing I for that dimension. As usual, the details turn out to matter, so let's write out the CART algorithm a bit more formally:

   Step 1: For each dimension of the data, choose the cutoff that maximizes the purity of the training data.

   Step 2: Choose the dimension that did the best in step 1.

   Step 3: Repeat steps 1 and 2 until there is no cutoff that improves the purity more than some threshold.

There are two key issues with CART (and decision trees in general). First, the tree has to be built up one step at a time. It's easy to think of examples where there are several good choices of which dimension to choose for the first step, but later on it becomes clear that one of the dimensions would have been better. Unfortunately, once the first choice is made by the algorithm, it can't go back. Thus, CART is a "greedy" algorithm—it maximizes the objective function in each step, but doesn't necessarily find the global optimum. The second problem with decision trees is overfitting— in general we don't know how big an improvement to the impurity we need to make in order to justify adding an extra level to the hierarchy. In the traditional decision trees implementations, this is addressed using a so-called "pruning" step where cross-validation is used to remove lower levels of the hierarchy that do not improve the classification accuracy on the test data. Pruning of the lower levels of the tree is straightforward, but for complex trees with multiple levels, overfitting may occur at multiple levels and be difficult to prune.

11.5 RANDOM FORESTS AND ENSEMBLE CLASSIFIERS: THE WISDOM OF THE CROWD

In practice, a much more powerful way to use decision trees is as part of an ensemble classifier. The idea of ensemble classifiers is that rather than training a single very complicated (and probably overfit) model, we train a (possibly large) number of simpler classifiers. These simple classifiers are then combined (in a simple way) to produce the final prediction. Ensemble classifiers are typically the top performing classifiers in contemporary machine learning competitions because they can leverage the complementary advantages of many different approaches.

Random forests are a type of ensemble classifier based on decision trees. The idea is to train a large number (e.g., 100) of decision trees (the "forest") based on random subsets of the features. Because the tree training algorithm (such as CART) doesn't see all of the dimensions, each tree remains relatively simple (much fewer decision levels are included then if all the dimensions are included). This means that the individual trees are much less prone to overfitting. This strategy is effective at combating overfitting, but each one of the trees achieves much worse classification performance because it is not using all of the dimensions of the data.

This is where the magic of the ensemble classifier comes in: To classify new data, the random forest gives *all* of the trees a chance to predict the class of the new observation. These predictions are then summarized (usually simply by assigning the new data to the majority class among the predictions). Because all of the trees were trained on random subsets of the features, they are all looking at different parts of the new data, and therefore their predictions are somewhat independent. Allowing all of the classifiers to vote turns out to be a very effective classification strategy. Amazingly, random forests are thought to be so robust to overfitting that the user can increase the number of trees arbitrarily—only computational constraints limit the number of trees that are used in practice.

More generally, as long ensemble classifiers are combining the predictions from simpler classifiers that are somewhat independent, it is thought that including additional classifiers will always improve the accuracy of the final classifier.

11.6 MULTICLASS CLASSIFICATION

Throughout the discussion of classification so far, we have been assuming that there are two classes, positives and negatives. In reality, of course, there are situations where we have more than two classes (often referred to as multiclass classification). Some of the linear methods discussed here (e.g., logistic regression) generalize in a totally straightforward way to the multiclass case (see Exercises). For others (e.g., SVMs), it's less clear how well they will work with more than two classes.

A classic multiclass problem in bioinformatics is classification of 3-D protein folds based on amino acid sequence. Based on extensive inspection of known 3-D structures, proteins can be classified into one of more than a thousand "folds," including things like Globin-like, β-sandwich, and TIM-barrel (SCOP release 1.75, Andreeva et al. 2007). Having sequenced a genome and predicted protein sequences based on the genetic code, researches might wish to know which (if any) of the previously known 3-D folds a novel protein belongs in. You can see why this multiclass situation is much more difficult than the binary classification problems that we've been considering so far: to apply a probabilistic model to this problem, we need to train a model for each of the folds. In this problem, that means thousands of models, as opposed to just two in the binary classification setting.

One relatively straightforward (but inelegant) way to deal with the multiclass classification problem is to break down the problem into a series of binary decisions and train classifiers for each of them. The most typical way to do this is using the "one-versus-all" approach where a classifier is trained for each class using the observations of that class as the positives and the positives from all the other classes as negatives. For example, to train a classifier for β-sandwich proteins, an SVM could be trained to distinguish between the known β-sandwich proteins and all other classes. This yields a binary classifier for each type of positive. New examples are then tested against the classifiers for each class. Although this approach can work well in many cases, the main problem is that often more than one of the classifiers will identify the new example as a positive. In this case, the user doesn't know which class to assign it to. It's important to note that in some cases, assigning a new datapoint to multiple classes might not be a problem, in which case the one-versus-all approach might be just what you want. Indeed, since proteins can contain multiple protein domains, the Pfam database uses a series of thousands of Hidden Markov Models or HMMs, where each one can potentially identify a positive in a new query protein.

An alternative to the "one-versus-all" approach is the so-called "all-versus-all" approach. In this case, a binary classifier is trained to distinguish between each pair of classes. In the protein-fold example, this means that an SVM would be trained to distinguish β-sandwich proteins from Globin-like, a second SVM would be trained to distinguish β-sandwich from TIM-barrel, and so-forth for every pair of classes. A new example is then compared to all the classifiers and assigned to the class which the most pairwise classifiers found positive. In the ideal case, a new β-sandwich protein is correctly identified by all the β-sandwich versus other individual SVMs, and all the other SVMs choose randomly. This approach was used for classification of protein folds and produced greatly improved results relative to the "one-versus-all" approach (Ding and Dubchak 2001). However, the major drawback of this approach was that the number of binary classifiers needed scales as the number of pairs of classes, $\frac{1}{2}m(m-1)$, where m is the number of classes. This means that as the number of protein folds approached 1,000, the number of pairwise SVMs needed approached 50,000.

In practice, for current protein fold predictions, this problem is actually handled is using nearest neighbor-style approaches—nearest neighbor classification doesn't require a model for the classes, only a means to compare the new datapoint to the previous observations. Sophisticated bioinformatics pipelines are used identify the most similar proteins whose folds are known, and the new protein's fold is inferred based on those. The key step becomes finding computationally efficient ways to compare the new protein to the entire database of known structures. Modern protein fold predictors such as Phyre (Kelly and Sternberg 2009) rely on huge sequence and structure databases and large compute clusters—users must submit their proteins for analysis via the World Wide Web.

EXERCISES

1. K-means clustering is a popular clustering method. As discussed in Chapter 10, clustering can be thought of as unsupervised classification. Explain why standard K-means clustering can be thought of as "linear" unsupervised classification. Use what you learned in this chapter about high-dimensional distances/kernels to devise a "non-linear" unsupervised classification strategy based on K-means.

2. Another common purity measure for decision trees in the "information gain" $I = -\sum_k f_k \log f_k$ where the logarithm is usually base two, k indexes the classes (e.g., positive and negative) and f is the fraction of datapoints in the class. What are the similarities between this measure and the Gini impurity defined here? What are the differences? (*Hint*: Think about possible statistical interpretations.)

3. Derive the classification boundaries for multiclass logistic regression, also known as multinomial regression. (*Hint*: Use the notation for the categorical distribution I gave at the end of Chapter 4.)

REFERENCES AND FURTHER READING

Andreeva A, Howorth D, Chandonia JM, Brenner SE, Hubbard TJ, Chothia C, Murzin AG. (2008). Data growth and its impact on the SCOP database: New developments. *Nucleic Acids Res.* 36(Suppl 1): D419–D425.

Breiman L, Friedman J, Stone CJ, Olshen RA. (1984). *Classification and Regression Trees*, Boca Raton, FL.: CRC press.

Ding CH, Dubchak I. (April 2001). Multi-class protein fold recognition using support vector machines and neural networks. *Bioinformatics* 17(4):349–358.

Hinton GE, Dayan P, Frey BJ, Neal RM. (1995). The "wake-sleep" algorithm for unsupervised neural networks. *Science* 268(5214):1158–1161.

Kelley LA, Sternberg MJ. (2009). Protein structure prediction on the Web: A case study using the Phyre server. *Nat. Protoc.* 4(3):363–371.

Karatzoglou A, Smola A, Hornik K, Zeileis A. (2004). kernlab—An S4 Package for Kernel Methods in R. *J. Stat. Software* 11(9):1–20.

Krizhevsky A, Sutskever I, Hinton GE. (2012). ImageNet classification with deep convolutional neural networks. In *Advances in Neural Information Processing Systems*, pp. 1097–1105.

Le Cun Y, Boser B, Denker JS, Henderson D, Howard RE, Hubbard W, Jackel LD et al. (1990). Hand-written digit recognition with a back-propagation network. In *Advances in Neural Information Processing Systems*, pp. 396–404.

Poultney C, Chopra S, Cun YL. (2006). Efficient learning of sparse representations with an energy-based model. In *Advances in Neural Information Processing Systems*, pp. 1137–1144.

CHAPTER 12

Evaluating Classifiers

12.1 CLASSIFICATION PERFORMANCE STATISTICS IN THE IDEAL CLASSIFICATION SETUP

As we saw in Chapters 10 and 11, the machine learning community has already come up with many creative approaches to classification that can work in a wide variety of settings, so most of the time we can choose from what is already available and avoid inventing new classification methods. However, in many cases molecular biology experiments will yield new types of high-dimensional data with which we would like to train a classifier and use it on new experimental data. In general, we don't know in advance which classifiers (let alone which parameters) to use in order to obtain good performance. Therefore, it's of critical importance for molecular biologists to know how to train classifiers correctly and how to evaluate their performance.

In the ideal case (described in Chapter 10), the available data have been divided into three parts. The first part (training set) is used to estimate the parameters of the classification model by maximizing some objective function. The second part (validation set) is used to compare classifiers and choices of parameter that can't be determined simply based on the training data—for example, choosing the classification cutoff, regularization penalty, kernels, neighbourhood size, k. The third part (test set) is data that was not used at all during the training stages and parameter selection. The performance of the classifier is measured using this test set (also sometimes called unseen or held-out data). There are number of performance measures that can be calculated, and we will discuss several in turn.

12.2 MEASURES OF CLASSIFICATION PERFORMANCE

Perhaps the most obvious and familiar of performance measures is the accuracy. This simply measures the number of correct predictions (or classifications) as a fraction of the total number of predictions. In practice, however, whether the accuracy is the most useful measure of classification performance depends on the problem. For example, consider a sniffer dog that has fallen asleep at the airport and lets all the suitcases through. If the number of illegal substances in the suitcases is very small say 1% of the suitcases, the sleeping dog will still be a very accurate classifier: The only mistakes will be on the suitcases with illegal substances (1%). The sleeping dog's accuracy will be 99%, but I hope it's clear that this is not the kind of behavior we want. In this case, we might be willing to accept a few mistakes on the suitcases without illegal substances, if the dog is very accurate on the small fraction of suitcases that actually have something bad inside. We might accept a lower accuracy overall, if we could be confident we were catching more of the illegal suitcases. This is a case where the "positives" (illegal-substance-containing suitcases) are rare, but more important than the "negatives" (legal suitcases). On the other hand, if positives and negatives are of similar importance and frequency, the accuracy of the classifier might be a good measure of performance.

My purpose in bringing up the example of rare, important positives is not purely pedagogical. In fact, in many molecular biology and genome biology applications, we are in just such a situation. For example, consider the classic BLAST (Altschul et al. 1990) homology detection problem: We seek to classify which sequences from the database are homologous to the query sequence. In this case, the number of homologous sequences (positives) is a tiny fraction of the database. Identifying a small number of homologues accurately is much more important than misidentifying sequences that are not homologous to the query.

In cases where positives are very important, as is typically the case in genome-scale molecular biology applications, people typically consider the "sensitivity" or "true positive rate (TPR)" or "recall." This is just the fraction of the positives that are out there that were successfully identified. I hope it's clear that the sleeping dog would get 0% in this measure (despite an accuracy of 99%). However, there is also an extreme case where the TPR is not a useful performance measure: the hyperactive sniffer dog. This dog simply barks constantly and classifies every suitcase as containing illegal substances. This classifier will achieve TPR of 100%, because it will classify the rare positives correctly as positives. However, as I'm sure you already figured out, the accuracy of this dog will be terrible—it will only get the positives right, and therefore the accuracy will be 1%, or 99 times worse than the sleeping dog. This is the bioinformatics equivalent of a sequence database search that predicts every sequence to be a homolog of the query. This is of no practical utility whatsover.

Thus, although it might be what we care about, on its own, TPR is not a very useful measure of classification performance. Luckily, TPR can be combined with other performance measures that improve its utility. One measure that TPR is often combined with is the "false positive rate (FPR)." The FPR is the number of truly negative datapoints that were (falsely) predicted to be positive. Notice how this will reign in the hyperactive sniffer dog—now we are keeping track of how often a positive prediction is made when there is no illegal substance in the suitcase. To fairly compare classifiers, we choose the parameters to produce an FPR that is acceptable to us (say 5% based on the validation set) and then ask which classifier has a better TPR on the test set. In the case of the airport suitcases, where only 1 in 100 suitcases are illegal, FPR of 5% corresponds to misidentifying five legal suitcases for every one that contains an illegal substance.

A low FPR is of critical importance for any classifier that is to be applied to large amounts of data: If you are trying to classify millions of data points and the true positives are very rare, even an FPR of 5% will leave you with thousands of false positives to sort through. For a BLAST search, 5% FPR would leave you with thousands of nonhomologous sequences to look at before you found the real homolog. As, I hope is clear from this example, the FPR is related to the false discovery rate (FDR) in the multiple testing problem that we discussed in Chapter 3.

Although the accuracy (ACC), FPR, and TPR are among the more widely used measures of classification performance, there are many other measures that are used. In some cases, different measures of performance might be more relevant than others.

SOME COMMON MEASURES FOR THE EVALUATION OF CLASSIFIERS

True positive (TP): An observation predicted to be in the positive class that is actually positive.

False positive (FP): An observation predicted to be in the positive class that is actually negative.

True negative (TN): An observation predicted to be in the negative class that is actually negative.

False negative (FN): An observation predicted to be in the negative class that is actually positive.

Accuracy (ACC): The fraction of predictions that are not errors (False positives and False negatives). $ACC = (TP + TN)/(TP + TN + FP + FN)$

True positive rate (TPR): The fraction of the true positives that are predicted correctly. $TPR = TP/(FN + TP)$

False positive rate (FPR): The fraction of negatives that are predicted to be positive. $FPR = FP/(TN + FP)$

Positive predictive value (PPV): The fraction of positive predictions that are actually positive. $PPV = TP/(TP + FP)$

Precision: Synonym for PPV

Recall and sensitivity: Synonyms for TPR

Specificity: Synonym for FPR

Unfortunately, comparing classification performance is not simply a matter of calculating the TPR and FPR on the test set and choosing the method that does the best. This is because the performance statistics (such as FPR and TPR) of a classifier can usually be controlled by changing a single parameter, known as the "cutoff" or "threshold." This is easiest to understand in the context of the linear classifiers based on probabilistic models (Figure 12.1). Using the notation in Chapter 10, where the features (or covariates) are represented as X, and the classes (or targets) are represented as Y (which can be 0 or 1), the goal of classification is to predict some new $Y_{n+1}, ..., Y_{n+m}$ given the parameters of the classifier (which were trained on the previous observations $Y_1, ..., Y_n, X_1, ..., X_n$) and $X_{n+1}, ..., X_{n+m}$. For example, since logistic regression directly models $P(Y|X)$, evaluating the classifier is equivalent to assessing how well the model of $P(Y|X)$ predicts some known Y. In the case of only two classes, I argued that the MAP rule says to assign the observation to the positive class if $P(Y|X) > 0.5$.

As is illustrated in Figure 12.1, if we choose a different posterior probability cutoff (instead of 0.5), we can obtain different classification statistics. We can usually find a cutoff that gives us the FPR or TPR that we want, but in general there is a trade-off between increasing the TPR and FPR. In order to compare classifiers that might be based on different classification

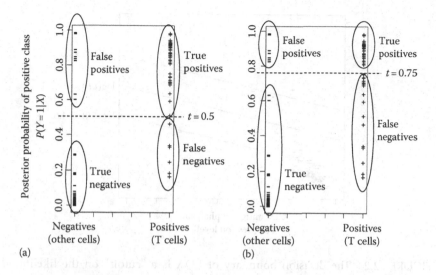

FIGURE 12.1 A logistic regression classifier as a probabilistic model for $P(Y|X)$. (a) The MAP (maximum a posteriori) classification rule (discussed in Chapter 10) for logistic regression implies that we should use a threshold (t) of 0.5 on the posterior probability to predict members of the positive class. (b) However, we can reduce the false positives by adopting a higher threshold of 0.75.

rules or models, we typically need to adjust the cutoff so that they are making predictions at the same FPR. Otherwise, it's trivial to find classifiers that achieve better performance with respect to one metric by reducing performance in another.

In the case of LDA or Naïve Bayes, changing the threshold can be interpreted as a change in the prior probabilities (see Exercises). In the feature space, changing the threshold for simple linear classifiers based on probabilistic models corresponds to translating the classification boundary, as illustrated in Figure 12.2. However, this is not generally true: For nonlinear classification boundaries, changing the threshold may have effects on the classification boundary that are not easy to predict.

12.3 ROC CURVES AND PRECISION–RECALL PLOTS

Thus, to "fairly" compare classifiers, we can compute the TPR as a function of the FPR as we vary the cutoff. We can visualize the classification performance as a plot of the TPR as a function of the FPR as we vary the cutoff. This is the so-called "ROC curve." For a perfect classifier, there will be some value of the cutoff where the positives are perfectly separated from the negatives. This cutoff will have TPR of 100% and an FPR of 0%

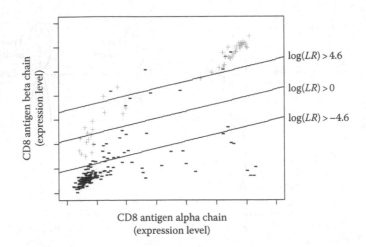

FIGURE 12.2 The decision boundary of LDA is a "cutoff" on the likelihood ratio. Changing the cutoff has the effect of moving the classification boundary. Positives (T cells) are represented as gray "+" and negatives (other cells) are represented as black "−."

and will fall at the top left corner of the plot. If such a cutoff exists, the ROC curve will look like a square. On the other hand, if the classifier has no ability to separate the positives from the negatives, for any set of predictions, the positives are simply drawn randomly from the positives and negatives in the data. This means that the FPR is expected to be equal to the TPR, and the ROC curve is a diagonal line. The ROC curve is the standard way to evaluate classifiers in biology, so if you have a classifier to evaluate and you don't know what else to do, make an ROC curve.

An ROC curve for the T-cell classification based on CD8 expression (data in Figure 12.2) is illustrated in Figure 12.3 for linear discriminant analysis (LDA) and an SVM with Gaussian kernel. You can see that both classifiers are performing much better than random guessing and are approaching perfect classification. The ROC curve indicates that the SVM with the Gaussian kernel outperforms LDA.

Very often people try to summarize the ROC curve with a single number, for example, by computing the area under the ROC curve (AUC). One attractive feature of the AUC is that the area under the curve for random guessing gives you an AUC of 0.5 and perfect classification gives you 1. The AUC allows you to make an absolute statement about whether your classifier is doing well or not, relative to a random expectation. On the

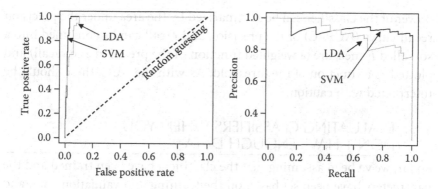

FIGURE 12.3 The ROC curve and P–R plot summarize the trade-off in the performance measures. Comparison of the classification performance of the SVM with a Gaussian kernel (black traces) to the performance of LDA (gray traces). On the left, the performance is compared using an ROC curve, while on the right the performance is compared using a P–R plot. Typically, the regions of most interest on these plots are the TPR at low FPR, or the precision at high levels of recall.

other hand, although in general, classifiers with greater AUC have better performance; by summarizing the performance this way, you can't get a sense of how the classifier will do on your problem: the AUC includes all levels of false positives, so you might have a classifier with a better AUC that actually performs much worse at the false positive level that is acceptable for your problem (see Exercises). In the simple example in Figure 12.3, although the SVM appears to outperform LDA at low FPR (top left of the ROC curve), both the methods have AUC measures near 1 (0.974 for LDA and 0.987 for the SVM) reflecting the fact that they are both classifying very well at all FPRs greater than 10%.

A possibly more useful measure of performance when positives are rare and important is the accuracy of the positive predictions. This measure is known as "positive predictive value (PPV)" or "precision." Since it's possible for a classifier to obtain a high PPV making few (but accurate) predictions, it's important to make sure differences in PPV are statistically significant, or compute the PPV as a function of TPR. Once again, there is usually a trade-off between these performance measures, so people tend to look at a so-called "precision–recall plot." An example of a P–R plot is shown in Figure 12.3. In this type of plot, a perfect classifier reaches the top right corner (PPV = 1 and TPR = 1). It's also possible to compute the expectation for random guessing, but it depends on the ratio of positives and negatives in the training set (see Exercises). Once again, the overall power of the classifier can be summarized by the area under the precision recall curve (AUPRC) or the precision and recall can be combined into a so-called *F*-measure (a weighted function of the precision and recall) and plotted as a function of the threshold. As with the AUC, these should be interpreted with caution.

12.4 EVALUATING CLASSIFIERS WHEN YOU DON'T HAVE ENOUGH DATA

So far, we've been assuming that the classifiers have been trained and the parameters have been set based on the training and validation data, and the performance measures (e.g., ROC curve) are computed on the test data that was never seen by the classifier before.

In practice, there are many cases when we simply don't have enough data to train and evaluate classifiers this way. Typically, when we are considering high-dimensional data, we need to estimate a few parameters for each dimension. We might try to model the correlation between dimensions to avoid getting distracted by features that are not really helping us. To do all of this, we need data. To train a classifier, a conservative rule of thumb is that you need 10 datapoints per parameter. You might be thinking—data, no problem—that's what genomics is all about. I've got thousands of measurements in every experiment. However, remember that in the classification world, there are parameters associated with *each class*, so we need sufficient observations from each class in order for the classifier to learn to recognize them. This becomes particularly problematic when the class we are trying to find is rare: say, we only know of 10 genes that are important for nervous system development and we want to identify more. Although we can use the "big data" to easily train parameters the genes that are not important for nervous system development (the negative class), we will also need enough known positive examples to train and evaluate the classifiers. If we use all 10 known genes for our training set, then we have nothing left to use in order to evaluate the classifier.

Amazingly, it is still possible to train a classifier even when you don't have enough data. The main way this is done is through cross-validation. (We have already seen cross-validation in the context of choosing penalties for regularized regression.) The idea of cross-validation is to do many iterations of the "ideal" classification set up, where you leave out a fraction of the data, train on the rest, and evaluate on the left out data. It's important to stress that classification performance should only be evaluated on held-out (or test) data. This is because the number of parameters in modern machine learning methods can be very large. If a classifier has enough complexity (e.g., enough parameters), it will be able to predict the training data perfectly. In general, we don't care how well a classifier can do on the training data—those are the suitcases where we already know what's inside. The classifier is usually only of interest for how well it can predict new examples. Although in the simple prediction problems used for illustration purposes here, it might seem easy to clearly delineate the training data from the test data, in complicated genomics and bioinformatics applications, there are typically many steps of data analysis. It's easy for (even very experienced) researchers to forget that some of the early steps of the analysis "saw" the entire dataset. In these cases, cross-validation performed on later steps of the analysis will overestimate the accuracy of the classification model (Yuan et al. 2007). In general, we have to be vigilant about separating unseen data and ensuring that it remains unseen throughout the data analysis.

To illustrate the magic of cross-validation, let's try to classify cells based on single-cell RNA expression data. As we saw in Chapter 2, the expression levels measured in these experiments fit very poorly to standard probability distributions. To simplify the problem somewhat, we'll start by trying to classify 96 LPS cells (positives) from 95 unstimulated cells (negatives) based on gene expression levels for the 100 genes with the highest average expression levels (over the entire dataset). We will start with twofold cross-validation, which means that we have randomly divided the data into two equal parts. We use one to estimate the parameters of the classifier, and then compute an ROC curve to evaluate the classifier on the part of the data that were left out.

Figure 12.4 illustrates ROC curves for twofold cross-validation. In each case, the ROC curve was computed on both the training and test data to highlight the importance of evaluating the classfier on the held out data. For LDA (Linear Discriminant Analysis, discussed in Chapter 10) and SVM (Support Vector Machines, discussed in Chapter 11), perfect classification is achieved on the training set in every case. However, for the test set, the classification accuracy is much lower. This illustrates how classification accuracy tends to be inflated for the training set, presumably due to overfitting.

If the difference in classification performance between the test and training set is really due to overfitting, it should be possible to reduce this difference using regularization (as discussed in Chapter 9). Indeed, Figure 12.4 shows that using penalized logistic regression greatly decreases the

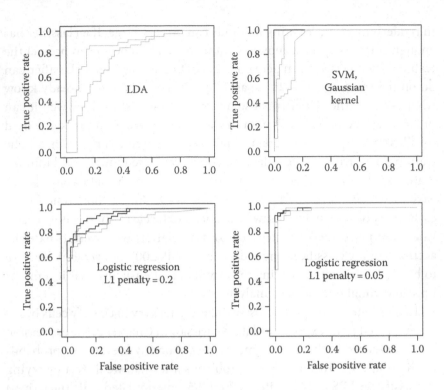

FIGURE 12.4 ROC curves on the training and test set for twofold cross-validation using various classification methods. In each panel, the performance on the training data is shown for the two random samples of the data in black traces, while the performance on the held-out samples is shown in the gray traces. Note that the classification performance depends on the particular random set selected in each cross-validation run, and is therefore a random variable. In the upper panels, no regularization is used and the classification models achieve perfect separation on the training set, but much worse performance on the test sets. In the bottom panels, regularization is used, and the training performance is similar to the test performance.

difference between the training and test sets. However, if the penalty chosen is too large, the classification performance is not as good on either the training or test set.

Note that in this example, I divided the data into two equal parts and we could look at the classification performance using ROC curves. In general, half of the data might not be enough to train the model, and you might want to use 90% of the data for training and leave only 10% for testing. In that case, you would want to repeat the analysis 10 times, so that each part of the data is used as the test set once. Since it's not easy to look at 10 ROC curves, instead you can pool all of the results from the left-out data together and make one ROC curve.

12.5 LEAVE-ONE-OUT CROSS-VALIDATION

In the previous example, I suggested dividing the training set into 2 or 10 fractions, and running the classification on each part. In the limit, you get "leave-one-out" cross-validation where the classifier is trained leaving out each datapoint alone. The classification performance measures are then computed based on the left out datapoints, and summarized at the end. I hope it's clear that leave-one-out cross-validation is making up for lack of data by increasing the computational burden: the parameters of the classifier are being re-estimated many times—proportional to the number of data points in the dataset. So, if a classification method needs a number of calculations proportional to the size of the dataset (say n) in order to estimate the parameters, the leave-one-out cross-validation estimate of performance therefore takes n^2 calculations. Nevertheless, even in today's "data-rich" molecular biology, we are usually data limited and not compute limited. So leave-one-out cross-validation is the most popular way to evaluate classification methods in molecular biology (and predictive machine-learning models in general). Because leave-one-out cross-validation uses almost the entire training set for each iteration, it is thought to give the most reliable estimate of the parameters, and therefore the best guess at how the classifier would perform on new data (if the whole training set was used). Figure 12.5 shows ROC curves for LDA and an SVM to classify cell type based on single-cell expression data from 100 highly expressed genes. Note that the classifiers both achieve very good performance, and there doesn't seem to be an advantage to the nonlinear classification using the SVM. This suggests that although the data is 100-dimensional, in that high-dimensional space, the classes are linearly separable.

A very important cautionary note about cross-validation is that it only ensures that the classifier is not overfitting to the data in the training sample. Thus, the cross-validation estimates of the classification performance will reflect the performance on unseen data, provided that data has the same underlying distribution as the training sample. In many cases, when we are dealing with state-of-the-art genomics data, the data are generated from new technologies that are still in development. Both technical and biological issues might make the experiment hard to repeat. If any aspect of the data changes between the training sample and the subsequent

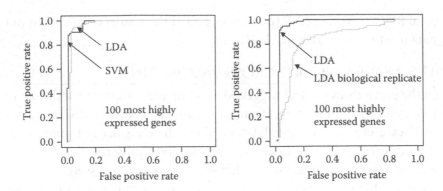

FIGURE 12.5 Leave-one-out cross-validation. In the left panel, the performance is shown for LDA and a support vector machine with a Gaussian kernel (SVM). The right panel shows leave-one-out cross-validation performance estimates on the original dataset (used for training and testing as in the left panel), and performance on a biological replicate where the model is trained on the original data and then applied to the replicate.

experiments, it is no longer guaranteed that the cross-validation accuracy will reflect the true classification accuracy on new data. In machine learning, this problem is known as "covariate shift" (Shimodaira 2000) to reflect the idea that the underlying feature space might change. Because the dimensionality of the feature spaces is large, and the distribution of the data in the space might be complicated, it's not easy to pinpoint what kinds of changes are happening.

I can illustrate this issue using the single-cell RNA-seq data introduced in Chapter 2 because this dataset includes a "replicate" set of 96 LPS cells and 96 unstimulated cells. These are true biological replicates, different cells, sequenced on different days from different mice. When I train the classifier on the original set and now apply it to these "replicate" data, the classification performance is not nearly as good as the leave-one-out cross-validation suggests it should be. ROC curves are shown in Figure 12.5. Note that this is not due to overfitting of the classifier (using a regularized model, such as penalized logistic regression does not help in this case). These are real differences (either biological or technical) between the two datasets, such that features associated with the cell class in one replicate are not associated in the same way in the second replicate. Because the problem is in a 100-dimensional space, it's not easy to figure out what exactly has changed.

12.6 BETTER CLASSIFICATION METHODS
VERSUS BETTER FEATURES

In the single-cell expression example given earlier, the SVM with Gaussian kernel (a sophisticated nonlinear classification method) did not perform any better than using LDA (an old-fashioned, overfitting-prone, linear classification method). Although at first this might seem surprising, it actually illustrates something that is reasonably common in biology. For very high-dimensional classification problems of the type we often encounter in molecular biology, the choice of features, and the ability to integrate different types of features into a single predictive model will usually be much more important than the choice of classifiers. For example, for the problem of identifying gene signatures, if you can make a good choice of genes, you won't need a very sophisticated classification method. In the simple case of the two-dimensional T-cell classifier, once we found better features (the T-cell receptor and one other gene in Chapter 10), we could classify almost perfectly with a linear boundary. For many complicated bioinformatics problems, choosing the right features turns out to be the key to getting good accuracy: Once you've found those features, simple classification methods can often do a good job. Currently, many important classification problems are thought to be limited by available features, not classification methods. For example, a major classification bottleneck in personalized medicine, identification of deleterious mutations or SNPs, appears to be limited by features, not classification methods. This means that in practice, feature selection and integration turns out to be the key step in high-dimensional classification problems.

To illustrate this let's return to our cell-type classification problem, based on single-cell sequence data, and now use the 1000 most highly expressed genes (instead of 100 as we were doing earlier). We can achieve essentially perfect classification (Figure 12.6). On the other hand, if we only had the three most highly expressed genes (remember, this is single-cell data, so measuring three genes' expression in single cells could still be nontrivial), we wouldn't be able to classify the cell types very well at all (Figure 12.6). In this case, using a better classification method (such as an SVM) does help us, but it's unlikely that we'll even get close to the accuracy we can get if we have much better features.

Indeed, there are many biological classification problems where the features we are using simply don't have enough information to do a good

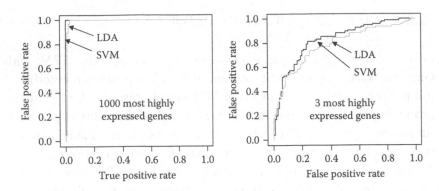

FIGURE 12.6 Better features make a much bigger difference than a better classifier. The two panels compare leave-one-out cross-validation classification performance of LDA (gray curve) to a support vector machine with a Gaussian kernel (SVM, black traces). On the left, the classifiers are trained with 1000 genes, while on the right they are trained with three genes. In both cases, the SVM performs slightly better than LDA. However, when data from 1000 genes is available, both classifiers can produce nearly perfect performance, while when data from only three genes is available neither classifier can perform well.

job. In these cases, trying lots of classification methods probably won't increase the accuracy very much—instead we need to be clever and find some new features, either by extracting new types of information from data we already have or by doing some new experiments.

EXERCISES

1. I said that the AUC of random guessing was 0.5. What is the AUC of the "sleeping sniffer dog" in the case where the data are truly 1% positives and 99% negatives? Why isn't it 0.5?

2. What does the P–R plot look like for random guessing if there are equal numbers of positives and negatives in the data?

3. Show that LDA with variable cutoffs corresponds to using the MAP classification rule with different assumptions about the prior probabilities.

4. In cases of limited data, one might expect leave-one-out cross-validation to produce better classification performance than, say, sixfold cross-validation. Why?

5. I have an idea: Instead of training the parameters of a naïve Bayes classifier by choosing the ML parameters for the Gaussian in each class, I will use the leave-one-out cross-validation AUC as my objective function and choose the parameters that maximize it. Am I cool?

REFERENCES AND FURTHER READING

Altschul S, Gish W, Miller W, Myers E, Lipman D. (1990). Basic local alignment search tool. *J. Mol. Biol.* 215(3):403–410.

Shimodaira, H. (2000). Improving predictive inference under covariate shift by weighting the log-likelihood function. *J. Stat. Plan. Infer.* 90(2):227–244.

Yuan Y, Guo L, Shen L, Liu JS. (2007). Predicting gene expression from sequence: A reexamination. *PLoS Comput. Biol.* 3(11):e243.

Index

Printed in the United States
by Baker & Taylor Publisher Services